VENUS
REVEALED

VENUS REVEALED

a new look below the clouds of

our mysterious twin planet

DAVID HARRY GRINSPOON

HELIX BOOKS

ADDISON-WESLEY
Reading, Massachusetts

Library of Congress Cataloging-in-Publication Data
Grinspoon, David Harry.
    Venus revealed : a new look below the clouds of our mysterious
  twin planet / David Harry Grinspoon.
      p. cm.
    Includes index.
    ISBN 0-201-40655-1
    ISBN 0-201-32839-9 (pbk.)
    1. Venus (Planet)   I. Title.
QB621.G75 1997   96-38448
523.4'2—dc20   CIP

Addison-Wesley is an imprint of Addison Wesley Longman, Inc.

Cover design by Robert Dietz
Text design by Karen Savary
Set in 11-point Minion by Pure Imaging Publishing, West Newton, MA

123456789-MA-0201009998
Second printing, June 1997
First paperback printing, March 1998

Find Helix Books on the World Wide Web at
http://www.aw.com/gb/

For Tory,
My evening and morning star

CONTENTS

ACKNOWLEDGMENTS

First I thank my wife, Tory Read. Without her love, patience, humor, wisdom, and killer editing, I don't think I could have survived the birth of this book. Some of the ideas found herein come out of conversations we have had and are really the product of our minds working together.

Mark Bullock helped in innumerable ways, including making the plots of Venus's motion in the sky found on pages 13 and 15, fact checking, and image gathering. I have also benefited from years of enjoyable conversations and scientific collaboration with Mark.

For critically reading all or part of the manuscript I thank Jake Bakalar, Jason Salzman, John Spencer, John Lewis, Nick Schneider, Jeff Moore, Buck Janes, Sue Smrekar, Lester and Betsy Grinspoon, Bill Hartmann, Tim Ferris, Glen Stewart, David Zuckerman, and Artie Rodgers.

Conversations or correspondence with the following people helped me in ways that turned up in this book: Tony Colaprete, Heidi Hammel, Peter Grinspoon, Josh Grinspoon, Carl Sagan, Antony Cooper, Carolyn Porco, Bruce Jakosky, Sasha Shulgin, Fred Whipple, Tom Donahue, David Crisp, Larry Esposito, Don Hunten, Fran Bagenal, Bill McKinnon, Kevin Zahnle, Josiah Carberry, Ann Sprague, Jim Head, Sasha Basilevsky, Mikhail Marov, Sean Solomon, Nori Namiki, Maribeth Price, Jim Pollack, Bob Strom, Damian Doyle, Joan Gabrielle, and Mike Kramer.

I thank Anthony Aveni for writing his inspiring books and for generously responding to my e-mail queries and suggesting source materials.

Geoff Skelton of the Fiske Planetarium provided time and planetarium access, which helped me in describing the movements of Venus.

The inimitable Carter Emmart drew the cartoons and illustrations found herein.

I thank the staff at Penny Lane in Boulder for the approximately 10^3 cups of coffee they made me while I was writing this.

I thank Jeffrey Robbins, Heather Mimnaugh, Jean Seal, Tiffany Cobb, Patricia Nelson, and Betty McManus of Addison-Wesley for their work on this project, and Jack Repcheck, formerly at Addison-Wesley, for helping get this book off the ground.

For helping get humanity off the ground, I thank the engineers and scientists involved in our first four decades of space exploration.

Finally, I thank my friends, family, and musical partners for putting up with me, or the lack of me, while I finished this book.

person, place, or thing?

Counting stars by candlelight all are dim but one is bright:
the spiral light of Venus rising first and shining best,
From the northwest corner of a brand-new crescent moon
crickets and cicadas sing a rare and different tune.
—ROBERT HUNTER, "Terrapin Station"

One fantastic California afternoon in late August 1989, a friend and I were driving the coast road to Neptune. We traveled south on Highway One, zooming and dipping along the contours of the Pacific coast. My fellow planetary postdoc Jeff Moore was at the wheel, so I had the ocean side and could watch the occasional whale surfacing above the white-caps to breathe. The stereo was cranked to compete with the sea breeze blowing through our open windows—B. B. King moaning about betrayed love. As we passed by a rocky cove I glimpsed, just for an instant, a beach full of a million pelicans all facing attentively in the same direction, as if listening to a sermon.

Even as we enjoyed another fine afternoon on Earth, our minds were focused 2.7 billion miles farther from the Sun, where *Voyager 2* was now only two days out from Neptune, and approaching at about eighty thousand miles per hour. We ourselves were barely able to do fifty on that twisty road and frequently had to slow down to a crawl behind an obese Winnebago, waiting for a straightaway to make our move. Highway One was flowing with tourists, and perhaps a few other pilgrims like ourselves heading to the Neptune encounter. Our destination was the Jet Propulsion Laboratory (JPL) in Pasadena, where, ever since Jupiter encounter in 1979, the faithful had gathered every time the *Voyagers* whipped through another planetary system. It was hard to believe that this would be the last encounter. But like the Wizard leaving Oz in his balloon, we had no way to turn *Voyager* back.

As we drove south and Earth spun ever eastward, the ocean rose toward the Sun. The haze on the horizon lit up with deepening colors, and thickening shadows highlighted the coast. It was one of those perfect days (except for the Winnebagos) when things just materialize when you need them: when the Sun was almost in the water a beach appeared just down the road for our sunset stop. As we sat in the sand and watched another day slip away for good we agreed that—although we couldn't wait to get to Neptune—all things considered, Earth must be the finest planet. What other place in the universe could have such landscapes, such music, such life?

Then, as we lingered and watched the dusk deepen, another world winked out of the twilight at us. It was "the spiral light of Venus, rising first and shining best." As darkness fell further, Venus became ever more brilliant until it looked as though it would burn a hole through the sky. Brighter than any star, never in the same place for long, Venus is a live wire, sparkling and dancing through our evening and morning skies.

We knew that somewhere out there was a spacecraft called *Magellan*, an emissary from Earth that was rounding the Sun for an arrival at Venus the following August. Although we had visited there many times with other robot craft, Venus was still such a mystery. The same planetwide covering of bright reflective clouds that lights up Venus for our viewing pleasure had frustrated efforts to see its surface. But *Magellan* could change all that. Its mission was to peel back the clouds with gentle radar fingers and reveal to us, at last, the face of our neighboring world. We knew that this would have to turn up some instructive and surprising things, but we had no idea. . . .

We lingered for a while, watching Venus descend through a darkening sky, following the Sun's footsteps into the night. It was hard to leave that beach and that sky, but we had to hit the road for our date with Neptune.

Over the next four days, Neptune and its moons completely blew our minds. But, as *Voyager 2* left the solar system to wander aimlessly among the stars, our planetary exploration program was in trouble. The *Voyagers* were launched in 1977, and as they traveled the length of the outer solar system, most funding for further exploration dried up. *Magellan* was the one remaining American interplanetary craft in space—the only one launched in a decade. So after Neptune our attention shifted toward the inner solar system as we anticipated *Magellan's* arrival at Venus—a place much closer to home in more ways than simple proximity. As Dorothy learned through her travels over the rainbow and we have learned through ours across the interplanetary void, there is no place like home. But there is one place that is, in many ways, not all that different.

For me Venus holds special fascination because of its many close connections to Earth, and the important role it has played in our changing ideas about the universe and our place in it. It is the brightest thing in our night sky, after the full moon, and has long attracted human worship, fear, and calculation. A slightly smaller twin to our planet in size and weight, Venus (we have recently learned) is also a currently active planet with a churning interior, young surface, and continually recycling atmosphere. It is in many ways the most Earthlike of other planets and a natural laboratory for studying some of the most crucial environmental tests facing us. We must explore and study the whole solar system in order to understand any part of it as well as we might. And we must explore other planetary systems, up close eventually, to really make sense of this one. But we may never find another place richer with lessons and insight about the home world than the world orbiting right next door to us—mysterious, cloudy Venus.

This book is about the human relationship with Venus and the way it has reflected and informed our changing relationship with Earth. In this story *Magellan* marks a new beginning, but it is only the latest part of an ancient adventure. Ever since we were separated at birth, 4.5 billion years ago, we have been on a long, strange journey back to Venus, our twin. I have tried to present some of the highlights of this trip. We pick up the story after most of it, in terms of the passage of time, has already happened, glossing over billions of years of planetary and biological evolution that allowed the appearance of human beings—one small, mobile part of the Earth's surface that developed brains, eyes, and curiosity, and started watching the sky. This telling of the story starts when humans first began looking up and noticing that one "star" was brighter than and behaved differently from all the others.

In Chapter 1, I describe the motions of Venus visible to the naked eye and discuss some of the interpretations and significance that various cultures have attributed to these. For at least thousands of years Venus has been a celestial object of worship and wonder. The classic Maya and the ancient Sumerians tracked its movements and told stories of Venus' underworld exploits in the service of humanity. Many early observers were quite sophisticated in their knowledge of these movements. For example, Mayan astronomer-priests knew about the subtle patterns of motion Venus repeats in our sky—tracing out the same figures five times every eight years. These result from a peculiar connection between the orbits of Venus and Earth that we still cannot explain. You might want to skip ahead briefly to the table and chart on pages 15 and 13, respectively, to find out what Venus is up to right now—if it is currently an "evening star" or "morning star" or too near the Sun for us to see. Then you can watch Venus's movements in the sky as you read this book and follow its path through our history.

In 1610 Galileo Galilei stuck some eyeglass lenses on the ends of an organ tube and pointed it at the sky. This remarkable invention, the telescope, allowed us to travel many times closer to the heavens without leaving the ground and forever changed our relationship with the planets, including our own. Galileo saw Venus traveling around the Sun, and showed that Earth, too, was merely a traveler, not the center. The wandering planets in the sky became worlds, and consequently our world was rudely stripped of its rank and forced to wander the solar system as one planet among many. Chapter 2 covers our hundreds of years of peering through these tubes, theorizing and fantasizing about what we saw. With telescopes we deduced that Venus was basically the same size as Earth, quite nearby, and shrouded in a thick atmosphere and enveloping clouds. Telescopes alone revealed little else. The clouds prevented us from learning anything about the solid body of the planet. But that didn't stop speculation—it encouraged it. We constructed an elaborate fantasy of a world very much like our own, probably inhabited and stuck in a more primitive stage of development, a living fossil of Earth's past, perhaps even the garden from which we had banished ourselves.

This dream persisted in various forms for roughly 350 years. In 1962 the first spacecraft from Earth, *Mariner 2*, reached Venus, inaugurating the era of planetary exploration and sending back shocking reports of a world not at all like home. Fueled by the technology of World War II and the paranoia of the cold war, the superpowers raced to send their surrogate missiles around the solar system and, for a while, scientists intent on studying the planets got free rides. More spacecraft were sent to Venus than to any other planet. In Chapter 3 I describe Venus exploration during the decades-old space age, when we started to go there ourselves, through our robot emissaries, and demand some answers.

Spacecraft exploration gave us an explosion of knowledge that created the new discipline of comparative planetology. Planets were no longer the province of astronomy. We came to know them as worlds with complex surfaces, atmospheres, clouds, and interiors. We began to address the problems of Earth and planetary evolution in a comparative sense. In Chapter 4 I summarize some recent ideas from this new field. One theme here is our growing appreciation of the important role of fluke events and historical contingency in determining the nature of planets. We have learned that to some extent planets, like people, are individuals whose personalities arise from their histories, and so cannot be predicted from deterministic laws. Because Venus and Earth are so close together and similar in size, the same age, and apparently made out of nearly identical materials, they provide a planetary twin study that may help us unravel the importance of "nature" (initial conditions) versus "nurture" (subsequent individual experience) in controlling the fates of worlds.

By the end of the 1980s, as spacecraft exploration—at least the first impulsive wave of it—wound down, we had learned much about the nature of our sister world's atmosphere. But we knew precious little of the surface. The clouds had proven an effective screen against our curiosity, even with our growing prowess and confidence as a species able to cross the distances between planets and ask direct questions. Then in 1990 *Magellan* arrived carrying our radar eyes, initiating a completely new era of discovery. In Chapter 5 I describe the history and findings of this spacecraft. From 1990 to 1994 *Magellan* circled Venus, revealing its previously cloud-hidden surface in stunning detail. We discovered many places of striking similarity to parts of Earth and others that defied all expectations and attempts at explanation. We see a world of giant volcanoes, rivers carved by flowing rock, steep valleys, rolling mountains, vast plains, and some kinds of places we have never seen before, for which we've had to invent new names. Perhaps most significantly, we find that Venus is a planet with ongoing geological processes—a recently active surface and recycling atmosphere. Unlike all other planets of the inner solar system, it's not dead yet. This makes Venus a great foil for learning about Earth.

The most unusual thing about Earth is our infinitely prolific biosphere—a feature completely unique to this planet as far as we know. In Chapter 6 I discuss the possibility of life on Venus. Here I play the devil's advocate. Venus is usually dismissed as an abode for life because it is assumed that other life will be chemically similar to us and therefore require similar conditions. I argue that a more agnostic viewpoint is justified by our grand ignorance of what life is, how it may be manifest elsewhere, and what the properties of inhabited worlds might be. In that final chapter I also discuss the prospects for future human exploration and visitation of Venus.

My historical and scientific coverage is deliberately uneven. I've learned that when traveling in foreign countries I get more out of spending extended periods in a few regions, getting to know them a bit, than trying to evenly cover every place. The same applies to interplanetary travel, and I will take that approach here. I have not attempted an even or comprehensive treatment of the entire subject. Instead, I have tried to write a book that will help your imagination to be captured, as mine has been, by Venus. For each stage of thought and discovery I've picked highlights that illuminate the character of what people were thinking about Venus. For example, I've gone into detail about some interplanetary spacecraft and briefly mentioned others. If I were to describe each space mission, each discovery and debate, in equal detail, the result would inevitably be either very long or quite superficial. So I've been selective.

Many books about science attach a name and institution to every idea, as in "in 1996 Jane Shmoe and Joe Blow of the University of Tuscaloola proposed, in an article published in *Nature*, that pigs might once have had wings." Here, I generally don't name names of living scientists involved in current research because I prefer to emphasize the collective human experience of the planets. Sometimes the activity of the scientific community seems like one giant mind at work, with ideas and thoughts that pass through it over the years. Not that we always agree—this mind is often not made up. It is restless, curious, hungry for new information and ideas, and often puzzled, conflicted, or confused. It is this collective perspective that I wish to present. I want to put the ideas themselves in center stage, to focus on what we have learned, not on the achievements of individuals. In this spirit, some of my own published scientific ideas and research results, and those of my students, are blended into this soup of distributed credit. So when I say "We discovered this" or "we learned that," I am usually referring to this group mind. Also, many of the debates I discuss are ongoing. Planetary science, especially Venus science right now, is full of conflicting opinions and puzzling evidence that we have not yet worked into coherent explanatory schemes. Our new, post-*Magellan* Venusian worldview is a rapidly evolving work in progress. Our ideas are farther from equilibrium than those in many other sciences, and that's what makes our conferences sometimes volatile and this field so much fun. It's easier to tell the story if I don't have to assign names and credit to the "camps" in every argument. When I offer my own opinions and new ideas, or indulge in speculation, I try to make it obvious that I am doing so.

This book is written not for planetary scientists but for everyone else. I do go into detail about some of the scientific instruments, ideas, and debates (although probably not enough for some of you). If it's too much in places, I would suggest either throwing the book down in disgust or just skipping to the end of the section. The book is structured so that you can pick up the flow at the beginning of the next section and still follow the action. I have tried to write in accessible language, either avoiding jargon or explaining it. There are no equations in the text, although I refer to them descriptively at times. Occasionally I make more quantitative elaborations in footnotes. I was able to enlist several friends, both scientists and nonscientists, to read rough drafts of the manuscript and make suggestions. Scientists sometimes made comments like "You should differentiate between the thermal and elastic lithospheres," "Shouldn't you describe the hypsometric curves?" or "What about supercalifragilistic flux nodules?"—whereas my other friends would say things like "This stuff about magnetic fields sounds like a bunch of mumbo jumbo to me." I have tried harder to satisfy the needs of the "panel of nonexperts." Scientists can read the scientific literature, if they want more information. Nonscientists do not have this option.

When we play the game Twenty Questions, we begin by requesting a classification for the object of our guesswork: "Person, place, or thing?" we ask. Venus has spent some time in each of these categories. Once it was an animated presence, a person/god in the sky carrying omens and messages about the underworld and the afterlife. Science transformed it into more of a "thing," a physical object to be investigated, and less of a guide or bringer of omens. Through the telescope Venus became a place, another world, and this required a major reassessment of Earth's place in the scheme of things. We knew that there was a *there* there, but hidden beneath those bright unyielding clouds, it could be *anyplace*. With our modern investigations, culminating in *Magellan*, all of these possible worlds vanished and Venus was revealed to be a specific place with unanticipated character and qualities of its own—a world of very real volcanoes and valleys, clouds, heat, and diffuse red light. We have recently found that geologically Venus, like Earth, is alive and kicking, absorbing and exhaling gases, slowly regenerating its skin. Perhaps it has now regained just a bit of the animation that the telescope robbed from it four hundred years ago.

Like most nerds, I was drawn into my field (planetary science) simply because it was the coolest thing I could imagine doing. When I was a teenager, spacecraft were landing on Venus and Mars, and setting off to photograph the moons of the outer solar system *for the first time.* With the possible exception of rock music (the other reason I felt proud to be an American growing up in the post-Vietnam and -Watergate world), what could be more exciting? As an adult who is concerned with the state of our species and our world at what seems to be a critical time for both, I am drawn to seek the relevance between our continuing planetary exploration and the problems we are currently facing. In these days of global environmental threats and renewed Earth consciousness, there is both a widespread spiritual longing and a pragmatic need to better understand our planet. How could we possibly hope to decipher the birth and life history of Earth without developing a knowledge of planetary evolution that encompasses the stories of all the planets of the solar system? Venus turns out to be particularly instructive for this quest—in general because it is a world with many similarities to Earth, and more specifically because of its highly relevant examples for studying acid rain, global warming, and ozone depletion.

Increasingly, we view Earth as a complex web of delicate balances and intricate feedbacks. To understand, reconstruct, and predict her workings, we have to understand the way this web works. Some have called this "Gaia," and some just "the Earth system," but whatever you call it, having a similar system nearby to study would be most convenient for gaining perspective, and testing our ideas in an enlarged arena. Recently we have learned that Venus seems to also embody an active planetary system with many complex feedbacks among surface, atmosphere, climate, and clouds. Thus Venus may serve as a valuable companion to Earth as we learn to live on, and with, our world. This may ultimately be the most lasting and vital gift that Venus will give us, an avenue toward the planetary self-knowledge that just may help save us from ourselves.

David Harry Grinspoon*
Denver, Colorado
August 1996

*Visit me at HTTP://sunra.colorado.edu/david

venus before the telescope: goddess at the edge of night

If the doors of perception were cleansed, everything would appear
as it is, infinite.
— WILLIAM BLAKE

SOMETHING IN THE WAY SHE MOVES

The striking gleam of Venus hanging low and bright in our morning
and evening skies, outshining all but the Sun and Moon, has demanded
the attention of ancient and modern sky watchers, inspiring religious
and scientific wonder. As long as we have been human, Venus has led us
to worship, fear, sing, speculate, invent, model, and calculate. It is like
no other celestial object in brilliance, and in the strangeness and complexity of its motions. It is not surprising that this brightly shining,
strangely looping, disappearing and resurrecting object has been per-
sonified, deified, given personality and purpose. Watching the intense
gaze of Venus as it moves slowly, intricately, through the seasons and

years, you get the feeling that there is a mind at work, and certainly
there is. It is your own.

Venus first appears as a faint jewel in the evening, barely outshining the glare of dusk, setting quickly on the heels of the Sun. At first,
she steps gingerly into night, departing before darkness sweeps the sky.
Then each evening she rises slightly farther from the Sun, glowing
longer and brighter. After thirty weeks she reaches her "maximum
elongation," the farthest she gets from the Sun. Now, no tentative
intruder into the night, she beams into the blackness, lingering long
after the Sun has departed, reluctantly setting three hours later. At this
phase Venus is so brilliant that if you get to a dark place on a clear
evening you can see your surroundings dimly illuminated by Venus
light and even find your Venus shadow. After seeming to pause here for
a week or so, she reverses course, diving down for the next ten weeks to
catch the Sun, setting closer and closer to sunset until the planet is
again consumed by the Sun's glare, nine months after first appearing.
Venus hides in the solar brilliance for a week and then reemerges on
the other side of night, appearing just before sunrise to dance the same
routine through our morning skies, rising earlier each week until, three
hours ahead of the Sun, she heralds the coming day. Again she pauses
before plunging back toward the Sun, disappearing this time for about
eight weeks, only to reemerge in the west, peering again through the
evening glare, beginning another cycle.

Through countless millennia, for much longer than anyone has
been watching, at least from around here, Venus has been jitterbugging
with the Sun, rhythmically swinging back and forth between the
morning and evening skies of Earth, just as surely as Earth has been
skipping, seen or unseen, across the nighttime skies of Venus. Eventually, who knows where or when, we began to take notice.

Venus in the sky, burning through the twilight, is a dazzling presence, brighter than all other planets combined. She commands our
mornings and evenings, moving to her own unique rhythms, dancing
with the Moon as it passes. I have seen Venus hundreds of times in the
evening and morning and even once at high noon during a total
eclipse, yet every single time I am stunned by its radiance. Who would
have dreamed that one day we would send a fleet of robots, driven by
remote commands, flying across space at unimaginable speeds to
plunge right into that brilliance and report home on what lies inside?

This great brightness and complex, enigmatic behavior led many ancient cultures to a great interest in—sometimes an obsession with—Venus. The Maya of Central America, for instance, were adept at tracking and predicting the date and direction of Venus's first reappearance in the morning or evening—an event of great religious significance. Today we tend to regard the planets as part of the "mechanical universe." We have learned that we live on a planet that is a large spinning ball following Newtonian law in an endless orbit around our home star, and we know that the other planets are other balls obeying the same laws on orbits around our shared Sun. This picture provides a simple, elegant, and accurate explanation for the strange cyclic appearances, loops, and disappearances of Venus and the other planets in our night sky. But much more than that, it is a correct description of the physical layout and motions of our solar system. We know this, most viscerally, because we have "been there, done that." We have gone to the planets and found them exactly where we expected them to be. Viva the Copernican revolution!* Our physics works.

And yet, with the pride we take in our modern understanding of the dimensions and motions of the larger universe in which we live, let's not forget that the knowledge and insights of earlier generations and civilizations of sky watchers were impressive in their own right, and even surpassed our own in some ways. Before we discuss the Venus cosmology of some earlier societies, let's review our modern picture of the motions of the planets, the lights that wander.[†]

Have you ever been waiting in a train at a station and thought, for a moment, that the train next to you had begun moving backward, only to realize that the forward motion of your own train was causing this sensation? Or been sitting in a car at a stoplight and sensed the car next to you moving forward, but then realized that it was you rolling

---

*The acceptance of the Earth as not the center of the universe but as an astronomical body among others, spinning and traveling through space, is often referred to as the Copernican revolution, after Nicolaus Copernicus, the visionary Polish astronomer who advocated the heliocentric (Sun-centered) universe in the 1500s. Of course, we now know that the universe is no more Sun-centered than it is Earth-centered. The Sun is in orbit around the Milky Way galaxy, which is itself in motion. The profound philosophical implications of this reassessment of Earth's importance have not yet completely sunk in among the Earth's inhabitants four hundred years later. The revolution continues.

[†]The word *planet* comes from a Greek word meaning "wanderer."

backward? These common experiences illustrate how observing from a moving platform can alter our perceptions of motion. Similarly, our views of celestial motions are dominated by the motions of our spinning and orbiting platform, the Earth. The path of Venus is easy to understand with our modern Sun-centered model of the solar system. However, we must first "cleanse the doors of perception," understanding and thereby circumventing a few persistent optical illusions that make this motion seem more complex and mysterious than it is. After all, our senses and brains have adapted well for finding food, shelter, and companionship here on Earth, not for the (until recently) useless task of fathoming the heavens. So let's help ourselves out with a little visualization exercise to shed some "commonsense" perceptions that can only get in the way.

It's always best to start at the beginning. First, we must step off the Earth and watch it spin. Everything in the sky—Sun, Moon, stars, and planets—appears to "rise" and "set," because we are on the surface of a round world that is spinning in space. Of course, we all know this, have been weaned on it; but to really understand the picture I am about to draw, it helps to be able to feel it, to know it on a deeper, nonanalytical, gut level. So the next time you watch a sunset, try this. As you see the reddening Sun "going down" toward the horizon, use your mental powers to stop it in its tracks, and instead force the horizon to rise up and slowly obscure the Sun. With a little effort, using the Sun's position as a steady reference point, you can, like the fool on the hill, see and feel the world spinning 'round.* Now, after the Sun is blocked by your western horizon and the light fades, observe the night sky slipping by as you spin beneath it, and feel yourself still turning away from the Sun. At midnight, sense the Sun far below your feet, obscured only by the turning Earth on which you stand. Several hours later, as the sky in the east begins to brighten and you anticipate, and then observe, the Sun "rising," picture the horizon falling away, revealing the Sun in exactly the same place where you left it the previous evening. This is the hard part. Here the altered state of sleep deprivation may come in handy as you try to loosen up the synapses and integrate old knowledge into new perspective. It's easy enough to say that the Earth has spun around

---

*This activity is done at your own risk. The author and publishers of this book take no responsibility for injuries resulting from loss of equilibrium due to a sudden internalization of celestial dynamics.

overnight, but harder to really believe, on the level of a gut feeling, that west has become east. (We are used to thinking that "never the twain shall meet.") Nevertheless, as you turn your head around 180 degrees to look back at the rising Sun, you are looking in the exact same direction as you were at the moment when the western horizon rose up and swallowed it the night before.* If you can feel this, you have successfully internalized the Copernican revolution.

So much for Earth's spin on its axis. Now, let's think about orbits. Earth is, of course, on a nearly circular orbit around the Sun with a period of one year (by definition). Our distance from the Sun is 93 million miles, 150 million kilometers, 8 light-minutes, or 1 astronomical unit (by definition), whichever is easiest for you to picture. (Good luck!) Viewed from this position in the solar system, the motions of all other planets fall into two different classes. Those we call "superior" planets, meaning that they are farther from the Sun than we are, travel slowly across the whole night sky, moving eastward against the far more distant background stars from night to night. Mars, Jupiter, and Saturn are the ones bright enough to see with your naked eyes, and thus were known before telescopes were invented. Each of these worlds pauses near the antisunward position, where it is high overhead at midnight, and reverses direction for a few months. This "retrograde loop" occurs when we, in our orbit, pass one of these slower-moving, more distant worlds, just as a car in a slower lane appears to move backward against the trees on the horizon as you pass it. Each world executes its loop, bowing to Earth as we pass, and then resumes the slow trip from eastern to western horizon. Eventually each disappears at sunset as it travels behind the Sun, only to reappear a few weeks later in the morning, resuming a slow circle dance around the night sky.

Our view is quite different for the two "inferior" planets, Mercury and Venus, which, since they orbit closer to the Sun than we do, are always near it in our skies. These planets appear only as evening or morning stars. They are never seen in the middle of the night, when your side of Earth is turned away from the Sun. They travel across our

---

*Actually, it's not quite exactly the same direction. Since the Sun itself must travel around our entire sky as we orbit in 365 days, in 12 hours it has moved half of $\frac{1}{365}$ of a circle, or about half a degree, against the background of distant stars. Half a degree is roughly the Sun's angular diameter in our sky. This angle corresponds to roughly two minutes of Earth's spin.

daytime sky, unseen because of the intense scattered light of the nearby Sun.* We can usually see them only near the horizon, right after sunset if they are east of the Sun, or else just before sunrise. The closer a planet is to the Sun, the faster it orbits. This is not just because those planets are traveling on smaller, shorter loops. Closer planets, caught more deeply in the Sun's mighty gravitational spell, actually move faster in their orbits.[†] So, just as Earth keeps lapping the outer planets, Venus constantly passes Earth like a runner on an inside lane (see Figure 1.1).

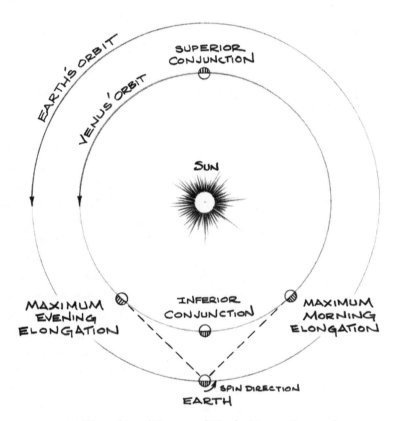

1.1 The orbits of Venus and Earth. (Carter Emmart)

*Venus is so bright that you can sometimes see it in the daytime if you know exactly where to look.

[†]The exact relationship is that the square of the orbital period is proportional to the cube of the average distance from the Sun. This is one of Kepler's laws of planetary motion. Kepler was an early follower of Copernicus, before it was popular or safe.

Now there is another persistent illusion, interfering with our intuitive comprehension of sky motions, that we should name and thereby help to transcend: the loss of the third dimension. The positional relationships and motions we observe in the sky certainly are three dimensional. The most distant thing we can see with the unaided eye, the Andromeda galaxy, is 5×10^{13}, or 50,000,000,000,000, times as far away as the Moon, which is the closest sky object (barring *Sputniks*, spooks, and space junk). But your stereo vision senses depth from the slight difference in direction between your left and right eyes, and everything in the sky is too far away for this to work. So we are stuck with a very powerful illusion that everything up there is the same distance away. We see the sky in only two dimensions—thus the "dome of night," or what astronomers call the celestial sphere. What we actually observe are the changing angles between objects on this imaginary sphere, and we see the planets seemingly wandering among the stars, which we have learned are actually millions of times farther away. We know now, thanks to the powerful insights provided by modern astronomy, that there is an incredibly large third dimension to these motions. But we simply do not see it under ordinary circumstances. This is one reason why a total solar eclipse is such a moving experience, provoking religious awe and fear through the ages. If you are fortunate enough to be in the right place and time for one of these spectacular events, you will see the Moon pass in front of the Sun and block its light, transforming your surroundings into sudden eerie darkness. At this moment, you do get a brief glimpse of the 3-D structure of the sky usually hidden from us by the limitations of our senses.

When we look at the motions of Venus relative to the Sun's position, what we see is the flat *projection* of Venus's 3-D movements on this illusory celestial sphere. The constant running back and forth between evening and morning skies, with disappearances in between, is simply the nearly edge-on view of an inside circular orbit. Running faster than we on her inside track, Venus is constantly overtaking and passing us. When we see her coming up in the evening, we are looking back over our left shoulder as she chases after us in our orbit, to overtake us at "inferior conjunction" when she disappears from our view, passing between us and the Sun. A week later when she reemerges into morning, she is already off and running, leaving us in her cosmic dust, receding toward the other side of the Sun (see Figure 1.2).

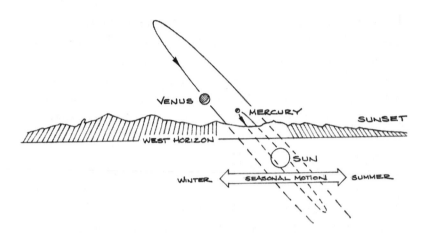

1.2 Looking toward the horizon at Venus's orbit projected on the sky. (Carter Emmart)

This cycle is easier to comprehend if you mentally stop the daily rotation of Earth and just watch Venus for a few years. If, as Earth spins on its axis, you fix your gaze on the Sun like a pirouetting dancer trying to maintain equilibrium, you can watch Venus swing from one side of the Sun to the other, bouncing back and forth between morning and evening. This may be easier said than done, but with my students in the planetarium I can dim the Sun and fix it steady in the sky, turn the time knob "up to eleven," and watch Venus yo-yo back and forth across the Sun as together they navigate yearly around the zodiac. Here the yo-yo string is the Sun's gravitational pull, which holds Venus in an orbit about three-quarters of the distance to Earth. In fact, if you do not have access to a planetarium, you can simulate this effect by getting a friend to swing a glow-in-the-dark yo-yo (or any luminous object on the end of a string) in a circle over her head, in the dark across the room from you. The yo-yo will appear to move back and forth in a line. So Venus seems to move as the Sun pulls it across the darkness of night.*

VENUS AT YOUR FINGERTIPS

You can easily measure the size of Venus's orbit around the Sun using your bare hands. In the yo-yo demonstration, it is obvious that the

---

*Seen from Mars or Jupiter, Earth is an inferior planet. Someday people living on Mars may see Earth swing between morning and evening star.

longer the string, the wider the angle that your simulated Venus will swing through. As Venus moves up and down through our twilight skies, the farthest it gets from the Sun is an angle of about 47 degrees. (This is "maximum elongation.") You can use your hand as a crude but effective sextant to measure angles on the sky. Hold it out along the horizon at arm's length with closed fingers, and its width will block out roughly 10 degrees.* When Venus is near maximum elongation, the extreme end of one swing of the yo-yo, you can measure about four hand spans from horizon to Venus at sunset.

Measuring this angle allows us to compare the size of Venus's orbit with Earth's. Picture two concentric orbital circles centered on the Sun, the outer one we ride on and the inner one that Venus follows (see Figure 1.1). As seen from our position on the outer circle, the angle from the Sun to the extreme points of the inner circle depends on the relative sizes of the two circles. The smaller the inner circle, the smaller the angle. A bit of trigonometry tells us that for a maximum elongation of 47 degrees the radius of the inner circle is about 70 percent of that of the outer circle.

If you don't like trigonometry or just want a more visceral confirmation, try this: Draw two large circles in the sand on a beach, or with chalk on a parking lot. Stand on the outer circle and see how many hand spans you can measure from the center (which might be your fire pit at the beach or your Camaro in the parking lot) to the edge of the inner circle. Experiment with the relative size of the two circles, and you will find that an angle of 45 degrees (four hand spans, give or take a few fingers) always corresponds to an inner circle with three-quarters the radius of the outer one. The angle depends only on the size ratio of the two circles. It doesn't matter whether they are one hundred feet across on the beach or 100 million miles across space. Your hands tell you that Venus is 70 percent as far from the Sun as Earth. The actual distance measured with more precise, if less intuitive, instruments is 0.723, on a scale where the radius of Earth's orbit equals one.† In more familiar units, Venus is 67 million miles from the Sun, as compared with our 93 million. Sunlight takes six minutes to reach Venus, and

---

*Your actual mileage may vary due to settling of contents. For reference, both the Sun and full Moon are one-half degree across.

†As I mentioned, this measurement, which defines the Earth-Sun distance as equal to one, is known as the astronomical unit (AU).

about eight to reach Earth. Thus, at its closest (inferior conjunction), Venus is about two light-minutes away from us. Armed (literally!) with this knowledge of relative sizes, positions, and motion, allow yourself to imagine the third dimension next time you view Venus in the sky.

FIVE AGAINST EIGHT

These are the roots of rhythm and the rhythm remains.
—PAUL SIMON, "Under African Skies"

Venus makes one trip around the Sun every 224 days. But since Earth is orbiting in the same direction, it takes a total of 584 days for Venus to go around and catch up with us again. This is how long it takes for Venus to come back to the same position relative to Earth (say, from one inferior conjunction to the next), so this is also the time period of the cycle of motions we see Venus make in our sky. This includes one complete evening appearance and one morning appearance, plus the two disappearance intervals.

There is something strange and wonderful about the numerology of this particular time interval. If you multiply 584 days by 5 you get 2,920. Divide this by 8 and you get 365 days, or one Earth year. Thus, there is a simple whole number ratio, a resonance, between the orbital periods of Earth and Venus. Venus completes five cycles, morning to evening and back again, in eight of our years. In musical terms, this is a polyrhythm, with the orbits of Venus and Earth beating five against eight.

Our solar system abounds with resonances, many of which we understand as resulting from repeated gravitational tugs pulling orbits into sync. Neptune and Pluto beat a slow but perfect three against two in their long circles through the outer reaches of our planetary system. Many of the icy Moons of Jupiter and Saturn are locked together into similar rhythms.

I think that these heavenly polyrhythms are the real music of the spheres. Many people, including Johannes Kepler, who first derived the quantitative laws of planetary motion, have long been obsessed with trying to find the "music of the spheres" described by Pythagoras. In keeping with the Western fascination with melody and relatively rigid sense of rhythm, this quest has typically involved efforts to fit measurements of planetary orbits into various melodic patterns. However, the more classical (in the sense of being older) musics of Africa with their refined polyrhythmic subtleties may be a better starting place to find the music in planetary orbits.* You can feel when you are "locked in," in resonance, in a circle of drummers playing parts complementary to yours. Venus and Earth have been locked in in this same way, beating five against eight, for billions of years.

There is no accepted physical explanation for the Venus-Earth five/eight polyrhythm. Some regard it as merely a coincidence, noting that the correspondence is not exact. (It is actually off by about 2 days out of 2,920.) But I suspect that this close rhythmic connection between the orbits of Venus and Earth goes back to the time when the planets were being formed from collisions between smaller "planetesimals." Resonances seem to have played a crucial (although poorly understood) role in this process. Computer models of alternate-reality solar systems have shown that the orbit of Venus becomes unstable if you remove Earth, so it seems that these two worlds may have been codependent from birth. The five/eight resonance may reflect a deep connection between the formation, locations, and motions of Venus and Earth that is still a mystery today.

Regardless of the reason for it, the five/eight resonance means that whatever Venus is doing tonight, perhaps shining high in the evening sky or making its first appearance in the morning before the Sun, you can be sure that it will be doing the exact same thing on this same calendar date eight years (and five cycles) from now.

So far I have described the reasons for, and patterns of, the east-west motions of Venus in our sky—the up-and-down motions as you look toward the horizon. There is also a north-south component to this dance (left and right, looking toward the horizon). Just as the Sun bobs north and south with the seasons, because of the tilt of Earth's axis, Venus follows suit. The place on the horizon where she first appears, the north-south path she takes, and the place where she finally disappears all change with the seasons. In terms of north-south motions, Venus anticipates the Sun by a few months in the evening and

*Of course, there are exceptions to any such gross generalization. Chopin's *Fantaisie-Impromptu* for piano, in which the right and left hands play 4 against 3, is a famous example of polyrhythm in the European tradition. However, it is a much more integral component of African music.

lags behind the Sun in the morning. This is because when you see Venus up high in the evening, she is east of the Sun, so you see her where the Sun will be a few months, down the road, in its slow annual trip around the sky. In the evening, Venus is farthest north in the spring and summer and farthest south in the fall and winter. When Venus emerges into our evening sky in winter, she first appears far to the south and then swings northward as she rises into spring. In the morning it's the opposite: We see Venus to the west, where the Sun was a few months earlier. In the morning, she is farthest north in summer and fall and farthest south in winter and spring.

Since Venus and Earth are out of sync on a yearly basis, this means that consecutive appearances always happen in different seasons. So, from one sky appearance to the next, Venus does not follow the same path away from, and back to, the Sun. Rather, she executes a different series of curves and loops to the left and right (as you look toward the sunrise or sunset horizon where Venus viewing takes place). But because of the five/eight resonance between the orbits of Venus and Earth described above, after five complete cycles (which takes eight Earth years) both Venus and Earth are back to the same positions in their orbits. Then Venus again rises on the same date and the motions start repeating. So there are five characteristic shapes that Venus repeatedly sketches on the dawn or dusk sky in consecutive appearances, and the same shape repeats exactly every five cycles, or eight years.*

When Venus dances and sways through another dawn and dusk, she moves up and down with the nineteen-month pace of Venus's orbit relative to our own. But her side-to-side swivel is provided by Earth's seasonal tilting as our gyroscopically fixed axis tilts toward and away from the Sun throughout our twelve-month orbit. John Lennon used to say that one of the secrets of rock 'n roll, was to get two interlocking rhythms going in the same song that each move dif-

---

*That the plane of Venus's orbit is slightly tilted from Earth's, by 3.5 degrees, adds a few added twists to these curves. This angle also explains why a *transit*, when Venus passes exactly between Earth and the Sun, and the dark outline of Venus can be seen passing across the Sun's bright disk, is a rare event observed, on average, only every fifty years or so. Even though Venus passes between us and the Sun at inferior conjunction every 584 days, it is almost always just a bit north or south of the Earth-Sun line and so does not ordinarily transit across the Sun.

ferent parts of your body, making dancing irresistible. Venus has been rocking and rolling through our skies for billions of years, with her orbit controlling the vertical rhythm and Earth's seasons controlling the horizontal.

Figure 1.3 shows the projected paths of Venus appearances in the evening for the next eight years, starting in 1997. Table 1.1 gives

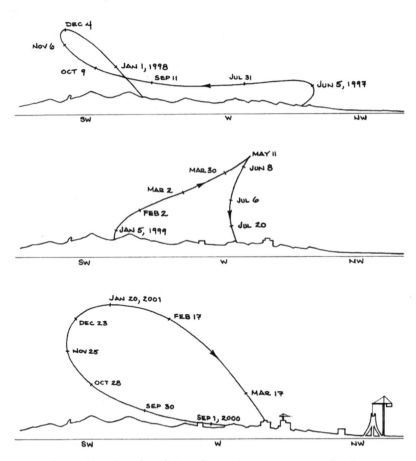

1.3 The projected paths of Venus's evening appearances for the next eight years. The paths of the morning appearances (which occur in between) are nearly mirror images of these. (Mark Bullock and Carter Emmart) Continued on next page.

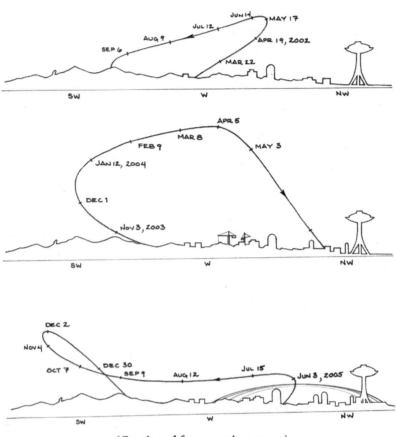

(Continued from previous page.)

the dates of first and last appearance, and best viewing dates for each of these. Note that eight years from the start of the table, after five complete cycles, the dates and shapes start repeating.

Now that you have had a chance to picture the 3-D motions of the planets as manifested on our 2-D sky, step outside and start watching. You can easily follow along as Venus traces its curves against the twilight. As you watch Venus in the evening, separating from and then rejoining the Sun, try to picture her on a circular orbit projected against the sky, descending to overtake us. Then watch her emerge weeks later high into the predawn sky and recede again, flying toward the other side of the Sun. With a little practice, you can start to get a feel for the size, shape, and pace of Venus's

1.1 OBSERVING GUIDE: VENUS APPEARANCES AND PHENOMENA 1997–2005
For an observer at mid-northern latitudes

YEAR	AS EVENING STAR (in west after sunset)	INFERIOR CONJUNCTION (date of closest approach to earth)	AS MORNING STAR (in east before sunrise)	VENUS NEAR MOON
1997	Late May to early-January 1998. Prominent after mid-October.			August 6: Venus 1.9° north of Moon. December 31: 1.3° south.
1998		January 16	Early February to early September. Highest in late February.	February 23: 1.6° north. March 24: 0.1° south. April 23: 0.1° north. May 22: 1.7° north.
1999	Early January to late July. Highest in mid-May.	August 19	Early September to late February 2000. Highest in late October.	January 19: 2.0° south. February 18: 1.9° north.
2000	Late September to mid-March 2001. Highest in late January.			February 2: 1.4° south. March 4: 0.6° north. November 29: 2.0° south. December 29: 1.8° north.
2001		March 27	Late April to mid-November. Highest in late July.	June 17: 1.7° north. July 17: 0.3° south. August 16: 1.9° South.
2002	Mid-March to late September. Highest in late May.	October 30	Mid-November to late June 2003. Highest in late December.	May 14: 0.8° north. June 13: 1.5° south.
2003	Early November to late May 2004. Highest in late March.			May 29: 0.1° south. November 25: 2.0° north.
2004		June 8 Venus transits directly in front of Sun today!	Early July to early January 2004. Highest in late August, early September.	April 23: 1.5° north. May 21: 0.3° south. November 10: 0.2° north.
2005	Late May to early January 2006. Prominent after mid-October. (This is a repeat of the 1997 appearance.)			August 8: 1.2° south. September 7: 0.6° north. October 7: 1.4° north. Nov. 5: 1.4° north.

orbit, and an enhanced awareness of the scale of our own endless circle around the Sun.

THE PIPER AT THE GATES OF DAWN

Long before we made telescopes and spacecraft, before we thought about orbits and spinning spherical worlds and the unique relationship between Earth, our home world, and Venus, the world next door, people everywhere knew that there was something peculiar about Venus. They worshiped, feared, and observed it, and learned to predict its motions. Early astronomers from many places and times were well aware of the resonance between the motions in the sky and the life-sustaining, and sometimes life-threatening, rhythms of the seasons, of growth and death, rain and drought. They noted and followed the lunar pulsations inherent in the rising and falling tides and were surely aware of the near-coincidence of this meter with the human menstrual cycle. It was obvious that there are intimate connections between the cyclic dramas of heaven and Earth.

It would be hard to overestimate the importance of the rhythmically changing sky for our pretechnological ancestors. For most of human history the sky was the calendar, the clock, and the compass. Now we look at the calendar to see what phase the Moon is in, when the tide will rise, and when the Sun will set. Weeks or months can go by for many of us without taking a good long look at the nighttime sky. In the past, without smog and lights, the sky was right there all the time and we looked at the Moon, the Sun, the stars, and the planets to find out what day, time, and season it was.

There are five bright starlike objects that move among the other stars in complex but ultimately predictable ways. Compared with the static stars, with their nightly clockwork spin and their slow repetitious seasonal parade across the sky, these bright wanderers are nonconformists, seeming to cruise the heavens with minds of their own. So, for any people prone to deifying and imparting personality to elements of the natural world, it made sense to assign important spiritual roles to the planets, the moving lights in the sky. And among these wanderers, it is not surprising that Venus has received the most attention and worship, for it is not only by far the brightest object in the night sky (except for the Moon), but moves in a way strikingly different from that of the other bright planets.* Venus must always have seemed a unique, animated entity. For our ancestors the details of the complex movements of Venus served as important harbingers of war and peace, feast and famine, pestilence and health. They learned to watch every nuance for the clues they could wrest of what nature had in store. They watched carefully, obsessively, through skies not yet dimmed by industrial haze and city lights, and they learned to predict accurately, for years and decades to come, the rising, setting, dimming, brightening, and looping of Venus.

Another important reason why Venus has always been singled out as an object of worship and wonder is the special relationship it enjoys with the most worshiped sky object of all: the heat-, light-, and life-giving Sun. Never far from the Sun in the sky, Venus is a small, bright sidekick, endlessly running circles round its larger companion. In the morning, Venus is the Sun's warm-up act. In the evening, Venus is an emcee taking the stage to exhort us to give it up one more time for the day's departing headliner, the Sun, before following him off into the wings. And over the months Venus repeatedly disappears into the glare of the Sun, only to be reliably reborn weeks later.

With behavior like that, it's no wonder that stories of birth, death, resurrection, and falling from grace are common to the Venus lore of many different cultures. The origin of the Judeo-Christian Devil as an angel fallen from heaven into the depths of hell is mirrored in the descent of Venus from shining morning star to the darkness below. This underworld demon, still feared today by people in many parts of the world, is also called Lucifer, which was originally a Latin name for Venus as a morning star. This is especially interesting in light of scientific discoveries we made centuries later, to be described in Chapter 3, which led to frequent comparisons between the environmental conditions we found on planet Venus and the biblical hell of fire and brimstone. In the ancient Middle East, Venus, called Ishtar, was worshiped as the goddess who evoked the power of the dawn. In ancient Sumeria

*Mercury has similar motions that repeat on a 116-day cycle, as compared with 584 days for Venus. But it is a much less commanding presence, often difficult to see, and it lacks any simple rhythmic orbital relationship with Earth. It has been said that Copernicus himself never actually saw Mercury. I've also heard that this is not really true, but even the fact that enough people could believe this tale to propagate it illustrates the rarity of a good glimpse of Mercury.

when Inanna/Venus descended after the Sun into *kur*, the "great below," it represented a dangerous and frightening abandonment that threatened sterility and extinction of terrestrial life if she did not return. For the Maya of Central America, Venus was the Sun's brother. When he preceded the Sun in rising from the underworld of night, he carried omens and messages for us from Brother Sun below.

An association with the underworld and the afterlife is quite common in Venus lore. Remember, the underworld had a very different meaning for people without a spherical-Earth worldview. This is where the Sun, Moon, and other celestial luminaries go each night between their evening exit from stage west and their morning entrance in the east, and it is to this same underworld that Venus departs during the one- and eight- week disappearance intervals that are a part of its familiar cycle. Think about it: If you believe, as common sense dictates, in a flat, immobile Earth, then where do those wandering lights go when they disappear from the evening sky and before they reappear in the morning? Is it the same place *we* go when we finish our life's travels? The implication is that there is a whole world of cosmic proportions that is contiguous with our own but inaccessible to us, except through the omens brought by these celestial messengers who, unlike us, can seemingly die and be reborn. In this reference frame the planets, which constantly disappear beyond the horizon and reappear, are seen as providing clues to knowledge of the underworld and hints about life after death.

As if all that weren't enough, another reason to worship Venus and find significance in her movements is that there are numerous connections between the timing of aspects of her motions and timescales of natural interest to humans. Perhaps most strikingly, the approximate 260-day length of a Venus appearance in the morning or evening coincides closely with the average length of the human gestation period.

The great beauty and striking presence of Venus led to an association by the Greeks with Aphrodite, goddess of beauty and love. Inanna, Ishtar, Astarte, and Venus are other names given to variations of this goddess in Western history, all associated with the planet. A knowledge of the close coincidence between the cycles of Venus and human pregnancy may have contributed to the persistent, but nonexclusive, Western attribution of female characteristics to Venus. The *Venus de Milo* and Botticelli's *Birth of Venus* (popularly known as *Venus on the Half Shell*) are icons of this imagery in Western culture.*

Nowhere in recorded history has an awareness of the short- and long-term Earth-sky polyrhythms been as advanced and integrated into cultural life as in the knowledge and beliefs of the ancient Mesoamericans, and in particular the Classic Maya of Central America, who flourished between A.D. 300 and 900. These Maya felt that we owed our very existence to Venus, whom they called Kukulcan, and their astronomer-priests repaid the debt with the blood of human sacrifice. Unfortunately, almost everything we know about the Maya's sophisticated and complex system of Venus observation/computation/prediction/worship comes from only four books that escaped the book-burning frenzy of the invading Christians. Included in these meticulously painted bark paper books is an abundance of astronomical information, including tables of solar and lunar motions and a Venus *ephemeris*, or table of motions, which is accurate for over a hundred years. The entire Mayan calendar, as were those of all Mesoamerican civilizations, was based on the 260-day Venus appearance interval. The 260-day Mayan calendar is still in use today in many areas of Guatemala.

The 260-day Venus interval and the 365-day year come into phase every 18,980 days, or 52 years. This time interval represents one Calendar Round, an interval of great ritualistic significance throughout Mesoamerica. The Maya were fascinated by the mathematical interweavings of different astronomically significant time intervals, and they celebrated the passing of a Calendar Round with lavish rituals of creation and renewal. The Mayan Venus Calendar even had a complex series of corrections to be applied on a range of timescales—the functional equivalent of our adding a day every leap year to keep our calendar in tune with the Sun. This resulted in an error of only two hours in five hundred years of elapsed time—all without abaci, slide rules, Macs, *or* PCs.

The Maya, and later the Aztecs, were keenly aware of the five different patterns that Venus repeats through our skies every eight years.

---

*This use of Venus to represent a certain ideal of femininity continues to the present. This Venus can be found, for example, in the title of self-help books on sexuality and in a popular game in which a refrigerator magnet of a naked *Venus on the Half Shell* may be dressed up in various magnetic outfits.

This is something that many modern astronomers know nothing about. The Mayan astronomer-priests had a different name and pictorial symbol for each of the five Venus manifestations (these are the designs that begin each chapter of this book). This is reflected in the *Popul Vuh*, the creation story of the Quiche Maya. Here, the two main characters are Venus and his twin brother, the Sun. In this story the cycle of Venus's appearances that we see in the sky reflects all of the key moments in the history of these people. His repeated risings symbolically represent creation and rebirth. The connection between the Venus cycle and the human gestation period may have enhanced the potency of this symbolism. The five major episodes of this story correspond with the five shapes Venus carves in the sky. In each, Venus and the Sun descend repeatedly to *Xibalba*, the Mayan underworld, to battle the evil Lords of Death, and prepare the world for the first dawn. The Lords of Death, with scary names like Pus Master, Stab Master, and Blood Gatherer,* are committed to the destruction of the human race. These are stories of intrigue and deceit.

In one remarkable episode Venus is entertaining the evil Lords with magic tricks. He wows them with a trick in which he kills his brother (the Sun), rips out his heart, and then commands him to come back to life. When brother Sun comes back alive, the Lords of Death are so delighted that they demand to have the trick performed again, on them. Clever Venus does as requested; only this time it is no trick. He kills them for real and goes home to claim victory. When Venus rises, a victorious morning star, he brings the coming of the world's first dawn. Some Maya say that if it weren't for these preemptive underworld strikes by Venus, the world today would be much more disease ridden. Each time now that Venus appears as a morning star to herald the arrival of his brother, the Sun, he reenacts the creation of the world.

The extreme gratitude owed to Venus for his leading role in creation was often repaid to him in human blood. Sacrifices in this cause were made in many parts of the world. In Polynesia and in North America the practice of offering up living humans to Venus continued into the nineteenth century. The Skidi Pawnee of Kansas and Oklahoma were diligent planetary observers who lived in close relationship

---

*If the *Popul Vuh* were released today as a gangsta rap record, it would definitely be required to wear a "parental advisory" sticker for its explicit violence!

with the sky. Venus plays a key role in their creation stories, and their debts to her were taken very seriously. Before dawn on April 22, 1838, a teenage girl name Haxti, who had been kidnapped from a neighboring Oglala Sioux tribe, was sacrificed to Morning Star at the first appearance of Venus, the culmination of a Skidi Pawnee Venus ritual that had lasted for weeks. This may have been the last of these ceremonies ever held. The moral indignation of those who did not understand the necessity of such rituals, to keep the world running and avoid disaster, proved too strong. And, of course, disaster did befall those who were forced to give up such practices. Their world came to an end so completely that historians must patch together these stories from fragments and fading memories.

For many cultures, the disappearance periods of Venus have been as important as the appearances. Of course, the disappearances are more noticeable if you are in daily contact with the sky, as the ancients in many places were. For those of us with a more fragmented and sporadic awareness of celestial events, the appearances call more attention to themselves than a specific length of absence does.

The Maya have always connected Venus with their god of rain. Associations of Venus symbols with rain symbols abound in their monuments, and in many stories Venus brings the coming of the rains. To the superficial modern observer this makes no literal sense, because as we watch Venus appear from one year to the next it is out of phase with our seasons, sometimes coming during rainy season and sometimes being absent then. But as Anthony Aveni has pointed out, there *is* a correlation between the length of disappearance of Venus and the season in which the disappearance and reemergence occurs. Because of the seasonal north-south weaving of Venus, described earlier, the disappearance intervals vary in length throughout the eight-year Venus cycle. The length of time when Venus is absent in between an evening and a morning appearance *averages* eight days, but it can vary from up to twenty days down to zero, and on rare occasions Venus can be spotted in the morning on the very next day after disappearing in the evening. The Mayan astronomer-priests knew that this variation in disappearance interval is seasonally correlated, and thus the length of time when Venus is gone can be used to predict wet and dry periods in Mexico!

The above is a good example of how the sky knowledge of some ancient astronomer-priests in some ways surpassed our own. Of course, it is true that this information is accessible to us and can easily be calculated in our computers from the well-determined basic elements of planetary orbits. But the Mayan astronomer-priests' awareness of exactly when and where on the horizon Venus would reappear and the subtle correlations between these appearances and the cycles of rain and drought, growth and harvest, on Earth, and the way in which this knowledge was integrated into religious and civic life is a lost art and science. This is related to a difference in perspective on where in the sky the action really is. We tend to look up, where the air is thin and clear, whereas the Maya and other early astronomer-priests looked over, at the horizon. Western astronomers have traditionally not been so interested in the horizon, but the ancient astronomy of the Mesoamericans and many other tropical peoples was horizon based, as befits followers of Venus, which can be glimpsed only near the day's edge.

The central position of Venus sightings in religious life and the horizon-based nature of their astronomy are both reflected in Mayan architecture. Carvings at Chichen Itza, Uxmal, and Copan celebrate Venus, and entire structures are designed and oriented to facilitate Venus-watching on important days in his cycle. In the eighth century, Copan was the site of a thriving Venus cult, which is memorialized with copious Venus symbols found throughout the surviving buildings. Temple 22 at Copan has a window through which you see Venus appear on important dates in the agricultural cycle. This window is not designed for the casual observer with no knowledge of Venusian pentameter. To make use of it, you have to know which of the five repeating motion cycles Venus is in at the time.

The Caracol of Chichen Itza is a building with many architectural features oriented for Venus viewing, including a round observatory dome on top with slots through which you can observe the position where Venus sets (pictured in color insert). By carefully watching where on the horizon Venus set on the day of his disappearance, the astropriests could accurately predict the number of days before he reappeared. Even the large-scale layout of Mayan architecture, including the relationship between buildings many miles apart, was often constructed with Venus in mind. The governor's house at Uxmal faces a large human-made mound four miles away. Venus rose over that mound on the morning when he made his first appearance in the southeast during one of his five paths.

The role of Venus in the lives of these cultures went beyond his being a central object of ritual and worship and a marker of events in mythical time. Venus played an important role in the unfolding of events in real historical time. At Bonampak, near Palenque on the Mexican Yucatan peninsula, are the ruins of a great temple where Venus symbols are found amid depictions of a great battle in which King Chaan-Muan, the ruler of Bonampak, emerged victorious. Chaan-Muan had a cadre of priests keeping secret Venus tables to predict important astronomical events. This allowed Chaan-Muan to act on these dates with the apparent complicity of the gods. On A.D. August 2, 792, Venus appeared in the morning after an eight-day absence, and on the same day the Sun passed directly overhead at noon. Chaan-Muan chose this day to launch a major battle against his enemies. The victory, commemorated at Bonampak, reaffirmed the divine, celestial source of his power.

As the god Quetzalcoatl, Venus also played a role in the defeat of the Aztecs by the invading Spaniards. Quetzalcoatl was identified with a myth, which historians think may be based on a real person, of an exile from the Toltec city of Tula in the tenth century. The light-skinned Quetzalcoatl was said to have traveled to the East, founded a new civilization, and predicted his own return. Unfortunately, the predicted date of his return coincided closely with the arrival of Cortés with his galleons and guns, in 1519. King Montezuma, guided by this myth, ordered his troops to put down their weapons, thinking that Quetzalcoatl/Venus had at last returned. The rest, as they say, is history.

Many of the monuments have been eaten by jungle, slowly eroded, or ceased to be aligned properly with the heavens because of the slow (twenty-six-thousand-year) precessional wobble of Earth's spin axis that even those great observers could not have accounted for. Although the books have been burned and the oral traditions largely lost, the lights in the sky, planetary and godlike, remain. Their constantly recurring motions are still following exactly the same paths as thousands of years ago, acting out the same stories of creation, birth, war, love, and death. These important records of our myths and origin stories are written in the sky and have not faded one bit, at least in those places left on Earth where the sky is still dark and clear. They remain there for our continued observation, wonderment, and perhaps even visitation. We may treasure them as a perfectly preserved link to some of the lost cultures of our human past.

MEN ARE FROM VENUS, WOMEN ARE FROM MARS

Much has been written about supposedly universal attributes of sky mythologies that transcend individual cultures. Some of this is suspect, but there are some good candidates for genuine near-universals. One of the best is the common association of the planet Mars with warfare. Mars *is* the red planet, we now know, because its surface is badly rusted. The associations with blood-spilling, warfare, and death are pretty straightforward and seem to have been made by many cultures independently.

Rapid movements, repeated highly visible appearances and disappearances, and a habit of dashing between evening and morning skies seem to have often given Venus a reputation for being capricious, proud, and willful. Similarly, the widespread tales of birth, death, resurrection, and underworld intrigue all mirror Venus's cyclic dance with the Sun.

The question of universal gender associations is much dicier. I have seen it stated many places that the planet Venus is associated with "feminine" qualities in appearance and motion, which are responsible for a widespread female designation. Both parts of this statement are dubious. Just what are these supposed feminine qualities? And is it really true that Venus has usually been seen as female? The popular British science writer Patrick Moore, in his 1959 book *The Planet Venus,* wrote: "A female association is in fact general, except in India; this is natural enough, since to the unaided eye Venus is the loveliest of the planets."

We find many comments like this in the literature, all written by men. It is simply not true that a female association is general. If you attempt any kind of cross-cultural reading of historical gender assignments for Venus, you find that he/she/it has been a real celestial gender bender that makes Madonna look tame in comparison. Venus has been male, female, and even sometimes oscillating, cross-dressing, or hermaphroditic.* Several cultures in the past observed a male morning star and a female evening star, or visa versa. For the Maya, Aztecs, and Toltecs, arguably the most

---

*Notice that even our term "hermaphrodite" contains the implicit assumption of Venus as female; the word melds male and female with Hermes (Mercury) representing the male and Aphrodite (Venus) representing the female side.

sophisticated and passionate Venus fans ever, Venus was mostly male. Now, personally, I do agree with Moore that Venus is the loveliest planet (but only if we disqualify Earth), but why should this necessarily lead to a female association? There are obviously some unquestioned stereotypes at work here. It is true that a female association was common to the Europeans who invented telescopes and their descendants who invented spacecraft and wrote most contemporary astronomy texts.

In many stories Venus is courted by suitors. This makes sense because Venus is tethered to the Sun, perpetually bobbing in and out of twilight but never entirely leaving it, whereas the other bright planets come and join Venus for a while, then wander off. But this contains no innate information about gender roles that transcend culture. Venus has been described as promiscuous and therefore female, but this is too obnoxious to warrant further discussion. However, the aforementioned near-coincidence of the length of a Venus appearance and the average length of a human pregnancy is perhaps more promising. Common associations with sexual love, fertility, birth, and new life all can claim some support from this. But there are just as many other Venus stories of war, death, disease, and macho conquests, so any claims for universal qualities or gender associations, written in the sky, are questionable. Obviously the planet has no gender but that which we impart.

At first I tried being completely gender-neutral in my writing, but this was unsatisfying because, to me, Venus is not just a "thing." Venus is not, in my mind, inanimate, and so "Cousin It" will never do to describe him ... or her. To make Venus an "it" is to reduce her to a dead lump of rock and metal surrounded by clouds and gases. He deserves better. Must we neuter Venus in order to avoid reinforcing dangerous stereotypes? I realized that just by using the name "Venus" I was accepting a cultural inheritance, with an implied gender association. Should I then randomly change names with which I refer to Quetzalcoatl, veering between Tai-Po, Ishtar and Hesperos, and occasionally back to Venus. This is a recipe for confusion. I decided to stick with the name Venus and its implicit female designation. Venus is female, a goddess, just as surely as Kukalkan and Mercury are male and Jupiter's moon Ganymede is named after his gay lover. Personally, I do think of Venus as female. I certainly acknowledge that this is contingent on my particular circumstances of birth, the result of my having been schooled and reared at a particular place and time. I know that if I were an Aztec living 500 years ago in Central Mexico, my Venus, Quetzalcoatl, would be a macho god.

So, I will often refer to Venus as female because that is the way that I, a product of my culture, think, and because it is better to have culture, personification and animation than none of the above. But I will throw in an occasional male reference as a reminder of all the lovely male Venus' that have and will continue to light up our horizons.

There is a modern convention of naming all surface features on Venus after female persons and deities. This has been justified as a kind of planetary affirmative action program, meant to rectify the male-dominated nomenclature found in the rest of the solar system. Yet, unfortunately it also results in a kind of ghettoization in which all of these fine goddesses and dead women are confined to the "separate-but-equal" world of Venus. It also reinforces and standardizes, for as long as our maps and computer files last, the idea of Venus as a feminine entity. Sometimes I wonder if this is a good idea, "naming" the features in the solar system in a way that makes a monument to the particular ideas of one culture. And surely Quetzalcoatl and Kukulcan deserve a place on Venus as much as anyone, despite their male gender. They were battling evil underworld demons in service to humanity centuries before there was an International Astronomical Union Committee on Nomenclature.*

Within this framework, however, the namers have done a great job, representing many cultures and individuals deserving interplanetary geographic immortality. What fool would dare quarrel with Artemis, Cleopatra and Mead?

VENUS IN THE CITY

Exit planet Venus for a Brooklyn stroll....
—DIGABLE PLANETS, "What Cool Breezes Do"

That many earlier societies had astronomer-priests, while today we have astronomers with priestlike knowledge of the depths of heaven, displays a common intuitive recognition, by humans with very different physical models of the universe, of the deep connections between the most basic facts of our existence here on Earth and the wider uni-

---

*This was, until quite recently, an all-male commmitee. A woman joined a few years ago.

verse. Sometimes it seems that in our current jargon-laden literature of the sky we could use a reminder of that sense of wonder and worship that was once more closely related to following the heavens.

For most of us, blinded by the lights and dazzled by technology's superficial control over nature, the movements of the planets have lost their central place in the rhythms of our lives and their power to predict the chaotic cycles of nature. Yet for those of us who choose to pay attention—and I hope that some of you may be inspired to join us—these motions still contain powerful messages, which have if anything become more essential because of the technological dimming of the celestial lights. Watch the movements of Venus over the coming months, feel the slow rhythm of her rising and falling, alternating every nine months between our evening and morning skies, a dance that beats in five/eight time to our earthly seasons and years. Notice the more subtle twists, turns, and loops she makes to the north and south as she repeats these steps, and her slow dimming and brightening in time to this pentameter.

To watch this motion over time is to see and feel yourself traveling through the solar system on a planet spinning on its axis and orbiting around the Sun, and to be aware of our closest companion in this journey. It requires a kind of patience, an attention span attuned to this rhythm with a tempo measured in months, which is a nice antidote to our media-clogged and traffic-frenzied daily realities.

There is nothing like the sky in a really dark place, far from city lights. But more of us live in cities all of the time, so this is where we get most of our views of the sky. Believe it or not, the urban sky has some advantages! Because only the brightest stars are visible in the city, we see a simplified night sky. This is good for beginners: it makes it easier to learn your way around. Since we usually identify the constellations by their few brightest stars, even experienced observers can become confused when confronted with a perfectly black sky shining with thousands of stars. If you live in a city, you can take advantage of the uncluttered sky to learn the brightest stars first. Then you can fill in the dots when you find yourself under darker skies. Once you have learned the major stellar landmarks and gotten used to their slow, steady seasonal march across the sky, you can more easily become aware of the nuances of motion of the bright planets that navigate through them.

Venus is an ideal object for urban astronomy—the best, really. Along with the Moon, the other naked-eye planets, and a few of the brightest stars, Venus has resisted the electric lighting-up of the sky that

has banished most astronomical objects from our cities and their surroundings. She is not the least bit intimidated by our streetlights and
tall buildings. When Venus is up high in the evening or morning, she is
a constant and brilliant presence even in well-lit places. Venus is easily
spotted during a concert or sporting event in an outdoor stadium, or
from a brightly lit parking lot. Viewed from the window of a city bus,
she skips along through the gaps in the skyline, hiding behind buildings
and jumping back out. When you start looking, she is always there.

THE EDGE OF NIGHT

For at least as long as recorded history, Venus has gotten under our skin
and moved us in different ways. I have discussed some of the specific
aspects of her appearance and motion that have contributed to this
attachment. But also consider this: Venus appears in our most colorful
and emotionally evocative skies. Not seen in the full black of midnight
or the full glare of day, Venus lingers at dusk or dawn, sometimes hanging off the corner of a crescent Moon or entertaining a lesser planetary
luminary. One of Earth's finest sights is of Venus alongside a new Moon
in the darkening evening, consorting among clouds of shifting colors
and forms. We see her at the fringes of the day, when a night breeze is
in the air, and a chorus of birds sings, while nature pauses and humans
reflect. As the Sun nears the horizon (or horizon rises toward standing
star . . .), its light must traverse a greater mass of air and suspended
droplets, dust, and smog. This atmospheric obstacle course scatters
more light, preferentially discarding blue and leaving a warm, ruddy
hue to the light at the day's dawning and dimming.*

The rich colors and half-darkness of twilight have a powerful
effect on the human psyche. There is an awakening of the emotions
and the imagination at twilight, as if the subconscious mind, having
been suppressed long enough during the day's rational activities under
the full glare of the Sun, is getting restless and starting to assert itself in
anticipation of its nightly reign. Or in the morning, when sleep has not
left our eyes and we are each emerging from our own personal underworld, when the world is still (except for those damn garbage trucks)

---

*This same effect dominates the quality of light seen at the surface of Venus
itself, which we will discuss later.

and pauses to breathe deep before the day, then we are vulnerable to Venus's spell. As the time of darkness and dreams approaches or recedes, there hangs Venus, peering out at us, shining with light reflected from beyond the edge of night (see Figure 1.4).

1.2 THIRTY NAMES FOR VENUS

PLACE OR LANGUAGE	NAME	ENGLISH TRANSLATION

Our world's cultures have spun and woven thousands of tales of Venus, and given her countless names. There is no way to be comprehensive here, since most names have been lost, but here is a brief sampler. I know there are a lot more to be found in books I have not read, and countless more never made it into any books.

PLACE OR LANGUAGE	NAME	ENGLISH TRANSLATION
Roman	Venus	goddess of love and beauty
Latin, ancient Roman	Lucifer	morning star, bringer of light
Latin, ancient Roman	Vesper	evening star, the west
Late Greek	Aphrodite	goddess of love and beauty
Early Greek	Phosphoros	dawn bearer
Early Greek	Hesperos	evening star
Chaldean, Sumerian	Ishtar	goddess of the east
Babylonian	Nabu	?
Sumerian	Inanna	queen of heaven
Egyptian	Quaiti	Evening star
Egyptian	Tioumoutiri	morning star
Hebrew	Helel	morning star
Arabic	Ruda	evening star
India, Sanskrit	Sukra	clear, bright
Chinese	T'ai-pe	beautiful white one
Gabon, West Africa	Chekechani	morning wife of the moon
Gabon, West Africa	Puikani	evening wife of the moon
Skidi Pawnee	?	white star woman
Skidi Pawnee	?	evening star
Aztec	Quetzalcoatl	feathered serpent
Aztec	Tlahuizcalpan-tecuhtli	lord of the dawn
Aztec	Xolotl	evening twin of Quetzalcoatl
Classic Maya	Kukulcan	Mayan version of Quetzalcoatl
Maya of Yucatan	noh ek	great star
Maya of Yucatan	sastal ek	bright star
Maya of Yucatan	ah ahzah cab	herald of the dawn
Quiche Maya of Guatemala	Hunahpu	morning star
Hawaiian	Hoku-loa	great star
Tahitian	Ta'urua	?
Kedang of Indonesia	Uno	evening star
Kedang of Indonesia	Lia	morning star

1.4 A waning crescent moon joined Venus in the morning sky, March 1995. (Travis A. Rector and Kamran Sahami)

2

venus through the telescope: earth's twin

In Ray Bradbury's 1951 science fiction story *The Long Rain*, a group of spaceshipwrecked Earthmen are forced to trek on foot through the choking jungles of Venus. There they are all driven mad by the incessant torrential rain which "... came by the pound and the ton, it hacked at the jungle and cut the trees like scissors and shaved the grass and tunneled the soil and molted the bushes. It shrank men's hands into the hands of wrinkled apes; it rained a solid glassy rain, and it never stopped." This Venus, an Earthlike, rain-soaked, heavily vegetated jungle world, was consistent with common scientific beliefs of the day. Today there is an irony to this imagery: we know that it is very nearly the opposite of what Venus turned out to be when we finally did figure out how to sniff about below the clouds. The most important single quality distinguishing Venus from Earth is the near total lack of water there. As much as it is anything else, Earth is the water planet; yet sister Venus is drier than bone. This fact is remarkable when we consider the numerous important similarities between Venus and Earth, and their birth together in the same litter of planets. It raises many questions of planetary nature and nurture, to which we will return. But first let's trace the origin and history of the popular vision of Venus as Earth's soggy twin.

The striking brilliance that makes Venus such a noticeable presence in our sky results paradoxically from the same feature that long kept her cloaked in mystery. The planet is completely shrouded in clouds that reflect nearly 80 percent of the Sun's light back into space, making Venus the brightest of worlds. Venus has surrounded herself with the most efficient planetary reflector known, and she bounces the Sun's intense light back into the darkness through which she orbits, like a cyclist with bright reflectors announcing her presence to the oncoming traffic. For millennia these luminous clouds have teased us, attracting our attention with their extraordinary glow, then thwarting our efforts to learn anything of the planet beneath them. This has fueled our imaginations with images of worlds that might be waiting below the impenetrable veil. Among the planets orbiting the Sun, none is closer to home, in distance as well as physique, than Venus. Yet she has been most elusive of all in giving up her secrets.

TALKIN' ABOUT A REVOLUTION

Some of us still believe the planets are gods, and many believe their motions influence human lives.* The shocking revelation of the Copernican revolution—that there is nothing central about Earth, nor by implication its inhabitants—was seen as so dangerous to the established order that it was branded illegal. Such knowledge threatened to turn the world upside down, or at least turn it loose from the moorings of biblically enshrined common sense. Understandably, this made the authorities very nervous and they tried, unsuccessfully, to stuff the genie of planetary consciousness back into their bottle of ignorance. The insight endured and forever changed the role of the planets in human affairs. It certainly has taken a while to sink in and gain mass acceptance, but by the early 1600s, after the rational, impassioned writing of Nicolaus Copernicus, the inspired numerological ravings of genius court astrologer Johannes Kepler, and finally the careful telescopic observations of Galileo Galilei, no informed, clear-thinking per-

---

*The tradition of the court astrologer persisted at least through the Reagan administration.

son could doubt that the Earth travels around the Sun and the planets are worlds like our own.

This band of Copernican revolutionaries was not the first to declare the world to be a planet, and the planets to be worlds, all circling the Sun. Ancient Greek scholars had known that the Earth is not flat but a sphere, and they believed the other planets to be spheres also. A few, notably Aristarchus of Samos, thought that Earth was not the center of the solar system, but the idea was heretical and never caught on. Instead the Greeks developed sophisticated and intricate Earth-centered schemes for the solar system (in this case, perhaps better called the terrestrial system) that did a good job of predicting and explaining the observed motions of the planets. These designs reached a peak in the complex, watchlike cycles-within-cycles, or epicycles, devised by Ptolemy at Alexandria in the second century A.D. His system was adopted by the Christian Church. Ptolemy's world was round, but it stood immobile at the center of the universe. And there it remained solidly anchored until, fourteen hundred years after his death, it was knocked adrift by the force of Galileo's telescope (see Figure 2.1).

Galileo did not invent the telescope, but on hearing, in 1609, that such an instrument had been made in the Netherlands, he quickly built his own device from an organ pipe and spectacle lenses. His initial motivation seems to have been for the telescope's intriguing military applications. (Those possessing a tool allowing them to scrutinize ships approaching in the distance would have an obvious

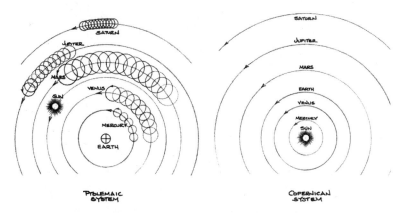

2.1 The Ptolemaic (Earth-centered) and Copernican (Sun-centered) schemes for the solar system. (Carter Emmart)

advantage.) However, within a few months he began to turn his crude devices toward the sky. Galileo was probably the first person to look through a telescope at Venus, or anywhere off Earth. If you have access to any kind of telescope, or even a good set of binoculars, you can look at Venus and see what got him into so much trouble with the church, and ultimately helped to ensure a successful revolution. Among the observations that put the final nails in the coffin of the geocentric worldview were the changing lunarlike *phases* of Venus. To our eyes Venus is a point of light that brightens and fades as it bobs around the Sun; a telescope reveals a spherical object whose phase, or illuminated portion, changes with its apparent distance from the Sun in the sky.*

ABOUT PHASE

Any sphere lit by a single distant source has a light half and a dark half. The phase you see depends on where you are looking from. It is determined by the amount of the light half you can see, which changes with your angle from the light source. It's easy to get a feel for this with a dark room, any ball, and a flashlight (see Figure 2.2).

The most well known phases, of course, are those exhibited by our own Moon in its familiar monthly cycle.[†] As it orbits Earth, we get a view from every angle, with respect to the Sun. We see a full moon rising at sunset when our backs are to the Sun, and shining overhead at midnight when the Sun is below our feet. The Moon appears as a half-illuminated sphere (called a quarter-moon) when we see it at a right angle to the Sun. We see a crescent moon when we are looking in a direction near the Sun and gazing mostly on the dark side. You will never see a crescent moon at midnight—only in the pastel skies of dusk (new moon) or dawn (old moon). Thus the celebrated association of Venus and the crescent moon, discussed in Chapter 1.

Earth seen from space also has phases. When you are watching a sunset or sunrise, you are right on the line separating the dark

---

*This does *not* require a dark or very clear sky. Early evening, when the surrounding sky is still blue and not too dim, is the best time to look. Individuals with exceptional sight may even be able to see the phases with naked eyes. Some pretelescopic references to the "horns of Venus," and horns on some Mayan Venus glyphs, have caused speculation that the crescent shape may have been observed by some long before Galileo.

[†]The words *month* and *moon* have the same root.

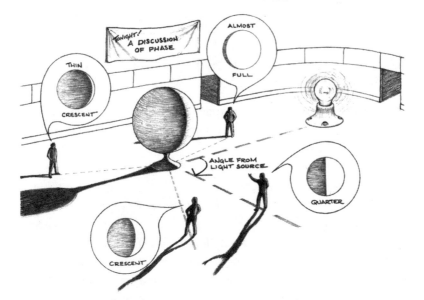

2.2 The phase you see depends on your angle from the light source. (Carter Emmart)*

and light halves of Earth. At that moment, if you could suddenly jump straight up a few thousand miles, you could turn around and see a quarter-Earth, half in sunlight and half in darkness.

As Galileo discovered, Venus also exhibits phases and therefore must be a round world shining by reflected sunlight.

Observing late in the year 1610, Galileo saw that Venus has phases of its own. Sometimes Venus is a crescent, sometimes half-illuminated, sometimes nearly full. He immediately realized the revolutionary value of this observation. Instead of publishing his finding in a straightforward manner, he chose to communicate it in a letter to Kepler as a puzzle, a Latin anagram,[†] that read:

Haec immatura, a me, iam frustra, leguntur - o.y.
["These things not ripe are read by me."]

---

*To see larger versions of these drawings see HTTP://sunra.colorado.edu/david

[†]This allowed him to establish that it was his discovery, without having to come out of the closet as a Copernican.

A rearrangement of the letters reveals the true message:

Cynthiae figuras aemulatur Mater Amorum
["The Mother of Love imitates the phases of Cynthia"; or "Venus imitates the phases of the Moon"]

The letters *o.y.* in the original message were set apart at the end because they did not fit in the anagram; Galileo was evidently a much better scientist than anagrammist.

Galileo's telescope also exposed the third dimension missing from our naked-eye sky, as revealed by the changing size of the disk of Venus. It appears to grow and shrink as it approaches us and recedes in its orbit. He saw Venus in the nearly full phase when it appeared close to the Sun and smallest in size. There was no escaping the implications: Venus was on the opposite side of the Sun from us, and shining by reflected sunlight. Copernicus must be right: the planets, including Earth, orbit the Sun. This observation pleased Galileo, who for years had been a secret fan of Copernicus. The church, however, was not amused by this claim that Earth, home of the "creation," was merely a small moving part of an immensely larger physical system, and not even the center of it all. Of course, they could have accepted Galileo's invitations to have a look through his telescope, and triumphantly declared that the Creator's work was even vaster than they had previously imagined (see Figure 2.3). Instead they condemned him and he was forced to recant his ideas or face death. He recanted to save his life, and was placed under house arrest until his death in 1642.* The Catholic Church finally pardoned him in 1992.

The word *telescope* refers to an optical instrument that can bring distant objects nearer, but an additional definition is found in most dictionaries. It also means to make or become shorter, compress or become compressed forcibly so as to occupy less space or time. In this other sense of the word, the new invention effectively telescoped our human egos, condensing our sense of ourselves into a much smaller space. This was the hard part to swallow, for the church and for our collective sense of place and identity.

---

*You can see an actual photograph of the middle finger of Galileo's right hand at http://www.jpl.nasa.gov/galileo/finger.html

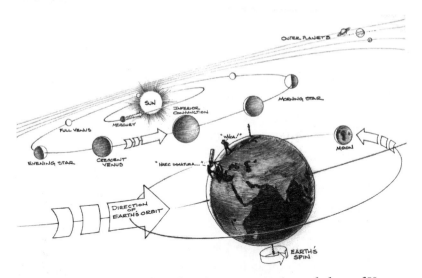

2.3 What Galileo saw: The changing apparent size and phase of Venus as it orbits the Sun are easy to see through a small telescope. (Carter Emmart)

Galileo lost his battle with authority, but his followers won the war. Over the last few centuries our relationship with the wandering lights in the sky has changed radically. Once merely animated celestial agents of destiny, planets have become other places, other worlds. Our relationship with our own planet has undergone a profound and closely related metamorphosis. It is no longer simply "the world," but one among many, whose uniqueness can be debated.* The technology and ideas of the Enlightenment and the industrial revolution have fed these transformations in worldview. No other invention affected our planetary self-image as profoundly as the telescope.

A PLURALITY OF WORLDS

There's a better world a-waiting in the sky, Lord, in the sky!
—FROM TRADITIONAL SPIRITUAL, "Let the Circle Be Unbroken"

---

*While the five planets bright enough to be seen with the naked eye each have hundreds of names, stories, and associations similar to those discussed for Venus in Chapter 1, the three planets discovered with telescopes each have one and only one name. These names—Uranus, Neptune, and Pluto—all come from the mythology of the Western culture that produced the telescope and first laid claim to distant space. In a sense, planet Earth was also discovered by telescope.

By the late 1600s many people were watching the planets with telescopes. There was no stopping the Copernicans; in fact, they went a bit
overboard. Many early telescopic observers believed quite literally that
the planets were other Earths and interpreted what they saw within this
framework. Once the commonsense, intuitive notion that the Earth is
flat, immobile, and the center of everything was discarded, then it no
longer seemed necessary to postulate anything special at all about our
planet. So the geocentric worldview was replaced by a widespread
belief in a *plurality of worlds*. If the planets are worlds like Earth, why
shouldn't the similarities extend to include the pervasive terrestrial features of life and civilization? In his *Anatomy of Melancholy*, published in
1621, the clergyman Robert Burton wrote: "If the earth move, it is a
planet and shines to them in the moon and to the other planetary
inhabitants as the moon and they do to us upon the Earth. But shine
she doth. . . . and then *per consequens* the rest of the planets are inhabited as well as the moon." It was no great mental leap to extend this
logic beyond the solar system. Later in the same essay Burton wrote:
"Then I say, the earth and they be planets alike. . . . If our world be
small in respect, why may we not suppose a plurality of worlds, those
infinite stars visible in the firmament to be so many suns with particular fixed centers, to have likewise their subordinate planets, as the Sun
hath his dancing still round him?"

   At first we knew virtually nothing about the other planets, except
for their startling, earthshaking resemblance to our own world in basic
form and motion. It was natural to fill in the blanks with familiar environments and inhabitants. I am not talking about a fringe element like
the "face on Mars" fanatics you may have encountered while catching
up on your tabloid reading in the supermarket checkout line. Many of
the most important scientists of the seventeenth, eighteenth, and nineteenth centuries endorsed the widely held view that the solar system
was full of Earthlike, inhabited worlds.

   One of the earliest telescopic observers was the Dutch scientist
Christiaan Huygens. Scientifically, Huygens was a heavy hitter. Among
other accomplishments, he deduced that Saturn had rings (and not
"ears" as described by Galileo), discovered a moon of Saturn, conceived
the wave theory of light, and made important contributions to the
physics of pendulums and rotating bodies. Captivated by the "plurality
of worlds" vision, Huygens came to believe in a widely inhabited solar
system. In his book *The Celestial Worlds Discovered, or Conjectures Con-*

cerning the Inhabitants, Plants and Productions of the Worlds in the Planets, published posthumously in 1698, he declared,

> We may mount from this dull Earth, and viewing it from on high, consider whether Nature has laid out all her cost and finery upon this small speck of dirt. . . . We shall be less apt to admire what this World calls great, shall nobly despise those Trifles the generality of Men set their Affections on, when we know that there are a multitude of such Earths inhabited and adorned as well as our own.

The two most important astronomical observers of the late 1700s were Johann Schroter, who has been called "the father of lunar and planetary astronomy," and William Herschel, discoverer of Uranus—the first "new" planet found with the telescope. Both believed passionately in life on the other planets. Herschel's belief in life on Venus was not based on pure fancy; it was based on observation and clever deduction mixed with fancy. He inferred correctly that Venus is so bright because of a global, reflecting cloud cover. He reasoned that clouds would help keep things comfortably cool on the surface, compensating for the more intense sunlight at Venus's location. Herschel was sure that there were intelligent beings living on Venus.

Perhaps this habit of extrapolating Earthlike environments into the heavens was so common because it fulfills a human need for a fantasy Earth somewhere else. The biblical heaven displaced by the Copernicans, or at least removed from its literal, physical home just beyond the outermost Ptolemaic sphere, was conveniently replaced by scientifically supported views of better worlds, or at least other worlds, waiting in the sky. The need once fulfilled by belief in the afterlife, in a place called heaven, in a connection to a world beyond Earthbound limitations, is now fulfilled for some by the idea of extraterrestrial life and intelligence.* This belief, if not scientifically proven, is at least consistent with current scientific beliefs. The hope for other nearby homes for "life as we know it" persists, but recedes towards distant star systems, or into the earliest history of our own solar system, chased through space and time by our swiftly expanding knowledge of the truly alien nature of other worlds.

---

*Present company included, I think. I do believe in extraterrestrials, and this brings me great comfort.

THE STEALTH PLANET

Early telescopic observations of Venus gave every indication of a nearby world much like Earth and only served to encourage these fantasies. But many of the similarities proved to be illusions produced by wishful thinking or poor optics abetting overactive imaginations. There are voluminous records of telescopic observations over the last few centuries reporting various details seen on Venus. Among the features "observed" on Venus are mountaintops, continents and oceans, polar caps and canals. The extensive literature includes meticulous sketches and maps of surface markings. There are detailed interpretations and speculations on the geology and biology and the nature of Venusian civilization.

Percival Lowell, the independently wealthy Bostonian astronomer who sold the world on his vision of Martian canals, also turned his telescope to Venus, where, beginning in 1896, he saw and carefully mapped an extensive, planetwide system of canals. He described these figures as quite distinct, and all his assistants swore they saw them too. Lowell felt that the canals of Venus appeared to be natural in origin, unlike the canals on Mars, which he claimed were artificial ones created by a civilization there. He thought that his canals would disprove the existence of clouds or a thick atmosphere on Venus, but the opposite would prove to be the case. Figure 2.4 shows Lowell's map of Venus.

Time, and improvements in telescope technology, were not on Lowell's side. The clouds are real, featureless, and global in extent. The weather never breaks on Venus. Seen through a good telescope in visible light, it is a bright, unblemished cue ball, white with a hint of yellow. The fact is, when it comes to using optical telescopes to learn anything definitive about what lies beneath the hazy covers, Venus is the stealth planet. You could not design a better defense against telescopic probing than the global Venusian* clouds.

---

*There was, for a while, a movement to stamp out the word *Venusian* and replace it with *Cytherian*. This was advocated in 1959 by Patrick Moore, who wrote: "There is no generally accepted adjective for Venus. 'Venusian' is common but ugly; 'Venerian' is even worse. 'Cytherian', an adjective derived from the old Sicilian name for Venus, is perhaps preferable, though not strictly correct." Personally, I like 'Venusian' and don't find it at all ugly. Venusian has made a strong comeback in the literature in recent years, and "Cytherian," though popular in the 60s and 70s, has faded.

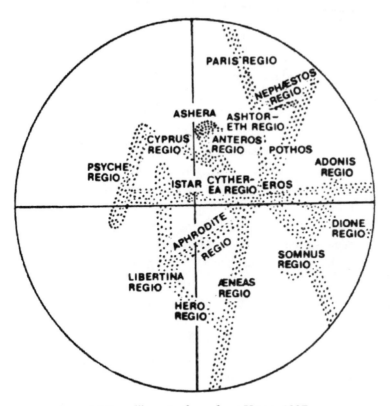

2.4 Lowell's map of canals on Venus, 1897.

Yet the literature on imaginary features seen on the blank disk of Venus remains as a cultural Rorschach test, revealing a lot about the biases, expectations, prejudices, and hopes of this world, even as it left the true nature of Venus's surface and evolution still shrouded in clouds. The bland, fuzzy, bright disk of Venus was a *tabula rasa*, a blank slate on which the wishes and expectations of the observers became inscribed.

There is another possible reason for the many "observations" of nonexistent features on Venus. Venus is remarkably featureless at wavelengths of light visible to us. Yet it is different when viewed at wavelengths slightly shorter than those our eyes can handle (in colors that are bluer than blue, bluer than violet). In the ultraviolet (UV), Venus is definitely not featureless. As F. E. Ross discovered in 1927, ultraviolet photographs show Venus awash in mysterious dark, high-contrast features (see Figure 2.5).* In the ensuing seventy years the physical characteristics and movements of these dark ultraviolet markings have been extensively studied and greatly clarified, but they remain mysterious to this day, and their source is one of the great unsolved puzzles of Venus, as we will discuss later.

Can we attribute at least some of the multitudinous reports of markings seen on the Venusian disk to these features in the UV? Some people are genetically endowed with the ability to see farther out toward the UV than others.[†] Could some of the reports have been accurate descriptions of features seen by observers with exceptionally good eyesight in the violet range? Probably not. The shapes and motions of the features described are generally very different from the wispy, rapidly rotating details actually seen in the UV.

Further doubt is inspired by experiments in which completely featureless balls are set up and "observed" through telescopes under simulated Venus-observing conditions. Many observers, including those with no knowledge of the astronomical literature, draw features remarkably similar to the "dusky shadings" reported on Venus. If you stare at a blank white wall or field of snow long enough, and try hard enough to see something there, you will. The same is true of a featureless planet seen through a telescope.

Many early observers reported a strange, faint glow coming from the night side of Venus. One might expect this "Ashen Light," as it has been called, to join the long list of observations that have been proved false by modern techniques. However, reports persist to the present. I have never seen it, but friends of mine have, and many careful Venus observers from the seventeenth century to the present seem sure that they have glimpsed it at one time or another. You have probably noticed how, when the Moon is a crescent, the unlit portion of it glows faintly. This "old moon in the young moon's arms" was first correctly explained by Leonardo da Vinci as coming from light reflected off of Earth onto the Moon. This won't work to explain the Ashen Light. Venus is just too far away for us to possibly see Earth light reflected off it. Some quite fanciful explanations have been offered up, such as

*Ross misinterpreted these dark features as places where breaks in the clouds revealed a dark surface below.

[†]Some other species, such as honeybees, have excellent eyesight in the ultraviolet.

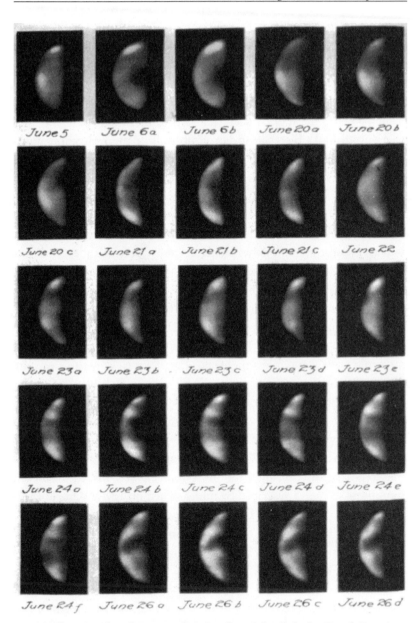

June 5 June 6a June 6b June 20a June 20b

June 20c June 21a June 21b June 21c June 22

June 23a June 23b June 23c June 23d June 23e

June 24a June 24b June 24c June 24d June 24e

June 24f June 26a June 26b June 26c June 26d

2.5 Photographs of Venus taken in ultraviolet light by Frank Ross in June 1927.

phosphorescent oceans or glowing cities beneath the clouds. Recent attempts to understand the Ashen Light have focused on possible optical effects in the atmosphere, like an extended twilight, lightning flashes, or an active aurora. No explanation has been agreed upon. Part of the problem is that the Ashen Light seems to come and go, and not everyone is convinced of its existence.

Telescopes are truly magical machines. These mirrored and lensed tubes, mechanically enhancing our senses, grab scarce photons of light, rearrange their trajectories, and funnel them into our eyes, bringing distant objects near. For almost all of the nearly four-hundred-year history of telescopes, viewers looked through an eyepiece and sketched or described what they saw. Our eyes and minds were an integral part of the observing and measuring equipment. No wonder the history of telescopic science has been so rich in entertaining dead ends. Eventually, photographic film replaced the more subjective human eye as the main light catcher at the end of the tubes, and now even film is becoming obsolete. The light pouring out of the magic tubes* encounters electronic detectors that convert it directly into a digital "data set," a grid of numbers to be reduced, processed, and analyzed. This may seem a few steps removed from the direct, visceral sensory experience of peering through an eyepiece high on a mountaintop on a cold, lonely, moonless night. But the pictures we get this way are often better than ever.

Before the invention of these somewhat more objective recorders of telescopic experience, even the most careful observers were often influenced by the reports of their predecessors. Take the case of Venus's moon. In 1645 Francesco Fontana, a Neapolitan lawyer and astronomer, reported discovering a moon of Venus. After this, the Venusian moon was seen and described by many skilled observers. One of these was Giovanni Domenico Cassini, an Italian-born astronomer who made many important discoveries including Saturn's moon Iapetus, which certainly does exist. These reports persisted for over a century and then died off. Venus has no moon.

The evolution of Venus's "observed" rotation rate is another illustration of the way preconceptions influence what we see. One property commonly derived from telescopic observations is the rate at which a body rotates on its axis. (This determines the length of a day you experience if you are there on the planet's surface.) To find the rotation rate of

---

*Actually, it's not *really* magic. It's all done with mirrors.

Venus, all you need to do is identify some feature or mark and note the time it takes to reappear, as it rotates around the globe. Beginning in 1666, Cassini made the first attempt to do this. He found a rotation period of 23 hours and 21 minutes. In 1789, Schroter refined this estimate to 23 hours, 21 minutes, and 19 seconds. At least eighty-six separate determinations of the Venusian rotation rate were published between 1666 and 1958, most of them clustering around a period of 24 hours. Remarkable that Venus should have a day length so close to Earth's—suspiciously close, in fact. We know now that nothing on Venus has a rotation rate even close to 24 hours.* This again demonstrates how even the most careful observers tended to confirm the results of others, and to find Earthlike characteristics in our neighboring planet.

Another group of rotation rates observed for Venus clusters around the much slower period of 225 Earth days. This period was first reported by the Italian astronomer Giovanni Virginio Schiaparelli in 1877, the same year he "discovered" the Martian canals that were aggressively promoted by Percival Lowell. Since this is exactly the same as the period of Venus's orbit around the Sun, this observation, if confirmed, would mean one hemisphere of Venus permanently faced the Sun. Our own Moon performs just such a "synchronous rotation" dance around Earth. Because of tidal forces from Earth's gravity, the Moon spins once on its axis in 29 days, exactly the time it takes to orbit Earth.[†] (As if it doesn't trust us, it never shows us its back.) After Schiaparelli's report, 225-day rotation periods briefly became all the rage and many other persuasive observers, including Lowell, reported similar results. Again we find no correlation whatsoever with any actual rotation rate measured by more modern techniques (see the next chapter). Moreover, *all* the early observers had Venus rotating in the wrong direction!

---

*Curiously, Mars has a day length of just over 24 hours. Its surface is visible from Earth and has many sharp features, so we knew its rotation period with an error of less than a second when we were still completely in the dark about that of Venus.

[†]For many years we also believed that Mercury exhibited precisely this behavior and had one very hot, permanently sunlit hemisphere and one permanently dark and freezing side. This idea was later shown to be wrong, but not before it had spawned some good science fiction yarns, such as Isaac Asimov's *Lucky Starr and the Big Sun of Mercury* and Larry Niven's *The Coldest Place.*

This history shows how the need to measure a value that simply cannot be determined with the available methods can lead to a false consensus that clusters near known values of quantities that are in some way related to the question at hand. The two most popular values for Venus's rotation rate were nearly identical, respectively, to Earth's rotation rate and Venus's orbital period. The intense desire for a number that made some sense, fit some pattern, or correlated with something familiar was so strong that it led some of the era's greatest scientists to agree on completely imagined "facts." This "bandwagon effect" (so called because of the easy comfort of forging consensus by jumping on someone else's bandwagon) is a danger in any era.*

It wasn't until the late 1950s, when radar technology became available for nonmilitary use, that we got some definitive answers about the rotation period of the solid planet beneath the cloudy smoke screen. Like much else about the planet below, it turned out to be completely different from what the early students of "Earth's twin" thought they knew but merely imagined.

WHO CARES?

At any point in my narrative of the science you should be able to stop me and say "Who cares?"† Why should we strain ourselves to collect such esoteric facts? Why should the taxpayers' money (or the king's ransom, during the era of many early discoveries) pay for this? And I had better have a good answer for you. Since you can't actually do so, this being a good old-fashioned dead-tree *book*, and not an interactive multimedia cyber-something, I'll have to do my best to anticipate such questions.

So, who cares about rotation rates? Why were all these smart people obsessed with figuring out how fast this distant object was spinning? The answer is that rotation rates help us to answer some of the major questions that we struggle with. We want to know "What's it like there?" and "How did all this come to pass?"

From the point of view of someone or something living on the planet being observed, the rotation rate determines the length of

---

*Except for this one, of course.

†Another good question, fair game at any time, is "How the hell do we know *that*!?" I always tell my students to feel free to ask me this at any point, and, believe me, they do.

day, how long you have between sunrise and sunset to go about your business on the surface of the planet. (Just reverse this if you are nocturnal.) Also, here on Earth, the rotation rate is firmly embedded in the rhythms of our biosphere. Our physiology and our cultural life follow the metronome of our planet's rotation. The macroscopic phosphorescent technological beast that is modern industrial society makes a standing wave of light that endlessly circles the globe, always facing away from the Sun.

On planets with atmospheres, like Earth, Venus, and Mars, the pace of changing solar illumination resulting from the rotation rate has a strong influence on the temperature, weather, climate, and seasons. It determines temperature changes on airless bodies like the Moon, Mercury, and asteroids.

Rotation also affects the *shape* of a planet. Centrifugal force stretches planets so they puff out at the waist, with a bigger bulge at the equator resulting from more rapid spin.* If we also have good observations of the shape, the extent of correlation between this equatorial bulge and the rate of spin can reveal the material properties of a planet's insides. The "squishier" it is inside, the more it will bulge in response to rotation. This in turn can help us figure out what the planet is made of at different depths, and how hot it is inside.

The spin also provides clues about the process by which the planets formed 4.5 billion years ago from collisions of smaller planetesimals. And, taken together, the spins of all the planets tell us more about how the solar system was formed.† This in turn can help us address the big-picture questions: "Is our solar system 'normal'?" "Are there others like us?" "Are we alone?"

EARTH'S TWIN SISTER

Although the mountains, continents, and canals so carefully sketched by early astronomers turned out to be imaginary, they did make many correct

---

*Earth's equatorial diameter is twenty-six miles greater than its polar diameter due to its twenty-four hour spin.

†Okay, okay, so we haven't exactly learned to decipher these clues yet, and the theory of the origin of planetary rotation is currently slightly confused, but we are *sure* this is all going to tell us something!

deductions about the physical nature of Venus, and these, too, conspired to reinforce the impression of another Earth waiting next door.

In 1761, at the University Observatory in Saint Petersburg, Russia, Mikhail Lomonosov observed a transit of Venus, a rare passage directly in front of the Sun. He was hoping to find Venus's diameter by measuring the size of its dark outline, which becomes visible during a transit in stark contrast to the bright Sun behind it. He was able to get a fairly good measurement and found that Venus's diameter was very similar to Earth's. But the precision of his measurement was hampered by a curious phenomenon. The edge of the Venusian disk, instead of appearing sharp, as he had expected, was fuzzy and indistinct, and a gray halo surrounded the planet. Consternation must have quickly turned to wonder as he realized the true source of this annoyance. Lomonosov had discovered the atmosphere of Venus. In his words, he had found evidence of "an atmosphere equal to, if not greater than, that which envelops our earthly sphere" (see Figure 2.6).

Observation and inference quickly gave way to speculative extrapolation. Fourteen hundred years before Lomonosov's discovery, Venus, as Kukulcan, had bravely battled underworld demons at the side of his brother, the Sun. For the next two centuries it was to act out a new, scientifically supported mythical role: Earth's twin sister.

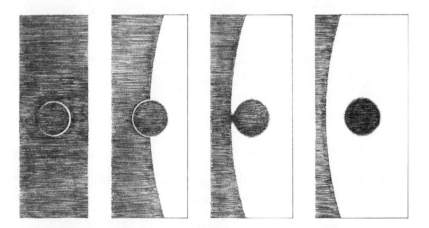

2.6 When Venus "transits" across the Sun, the presence of its atmosphere is revealed by several unusual optical effects: first a bright halo surrounds the planet as it nears the Sun, then a dark "umbilical" appears as it first passes in front of the bright star, and finally the edge of the disk appears fuzzy and indistinct. (Carter Emmart)

Considering what we knew about Venus at the time, this is not surprising, especially given the widespread post-Copernican propensity for finding familiar environments on the other planets. Lomonosov's discovery of the atmosphere, and the correct inference by many early observers that the bright, featureless appearance of Venus is due to a planetwide cloud cover, led to an irresistible line of reasoning. Venus is nearby and the same size as Earth. It has a thick atmosphere full of clouds—clouds, it was assumed from terrestrial experience, made of water. Since Venus is closer to the Sun, it must be somewhat hotter there than here. Therefore, Venus must be a hot, wet, thickly vegetated planet—a swamp world. Everything we actually knew about the planet was indeed remarkably like Earth, so it was easy to extrapolate this likeness to the surface of the planet and imagine a sister Earth beneath the clouds.

In his 1888 book *Astronomy with an Opera Glass,* Garett Serviss wrote

> It is an interesting reflection that in admiring the brilliancy of this splendid planet the light that produces so striking an effect upon our eyes has but a few minutes before traversed the atmosphere of a distant world, which, like our own air, may furnish the breath of life to millions of intelligent creatures, and vibrate with the music of tongues speaking languages as expressive as those of the earth.

In fact, Venus and Earth really *are* remarkably similar in many of their basic physical characteristics. In a system of nine planets where the smallest (Pluto) could fit inside the largest (Jupiter) 240,000 times, they are very nearly the same size and mass, with Earth weighing in as the slightly larger twin. Shrink Earth to the size of an official NBA basketball and, on the same scale, Venus is the size of a soccer ball.* When it comes to the basic property of size, Venus and Earth are very much in the same ballpark, and no other known planet is even in the same league. One thing that comparative planetology has taught us is that size is everything; it's the single most important quality for determining the course of a planet's evolution (more on this in Chapter 4).

---

*Actually, the planetary difference is slightly smaller; Venus would be 2 percent too large for official soccer play, probably not enough to be disqualified. On this scale, Mercury is a softball and the Moon a tennis ball. Mars is a big ripe red melon.

Also, if location has any influence on planetary character, there was all the more reason to expect a very familiar place. For Venus is (as was well known by early telescopic observers) literally right next door. The closest planet to Earth, at every inferior conjunction she swings to within one hundred times the Moon's distance.

CARBONIFEROUS PARK

Several scientists even speculated that Venus was undergoing an evolutionary phase similar to the Carboniferous period on Earth, over 200 million years ago, and was thus at a stage preceeding the development of "advanced life" like that found on Earth. The Carboniferous ended 100 million years before the start of the Jurassic era, which now enjoys Hollywood stardom rare for a geological period. It was a period of global warming on Earth, marked by high temperatures, a high sea level, and an abundance of swamps and inland seas. This was the time when decaying vegetation gave rise to the supply of fossil fuels, which we are now so impulsively burning.

Svante Arrhenius, the Swedish chemist who won a Nobel Prize and published numerous popular books, wrote in 1918 that "the humidity is probably about six times the average of that on Earth. We must conclude that everything on Venus is dripping wet. The vegetative processes are greatly accelerated by the high temperature; therefore, the lifetime of organisms is probably short."

A variation on this theme briefly became popular in the 1950s. Fred Hoyle, a famous British astrophysicist, proposed that the surface of Venus was covered in oil and that the clouds may be made of oil droplets. Observations going back to the 1930s had detected the infrared signature of carbon dioxide in the atmosphere.* This was consistent, Hoyle proposed, with a planetwide coating of petroleum. In 1955 he wrote, "Venus is probably endowed beyond the dreams of the richest Texas oil-king." This "Hoyle Oil," as it came to be called by astronomers, was in a sense consonant with the Carboniferous view of Venus, since that is the period when Earth's oil reserves were deposited. But, of course, Earth in the Carboniferous was covered not with

---

*This will be described in Chapter 3.

oil but with vegetation, which became oil only after millions of years underground.

As recently as 1955, Russian astronomer G. A. Tikhoff was publishing scientific papers with detailed inferences on the nature of Venusian vegetation based on current observations. Tikhoff wrote:

> Now already we can say a few things about the vegetation of Venus. Owing to the high temperature on this planet, the plants must reflect all the heat rays, of which those visible to the eye are the rays from red to green inclusive. This gives the plants a yellow hue. In addition, the plants must radiate red rays. With the yellow, this gives them an orange color. Our conclusions concerning the color of vegetation on Venus find certain confirmation in the observation . . . that in those parts of Venus where the Sun's rays possibly penetrate the clouds to be reflected by the planet's surface, there is a surplus of yellow and red rays.

This portrait of Venus, as a verdant, rainy, overgrown swamp planet—perhaps complete with tree ferns and jungle animals—became widespread in the popular and scientific literature throughout the nineteenth and most of the twentieth centuries. This vision also fit into a larger view of solar system evolution that remained popular for centuries. According to this scheme, the planets farther out from the Sun were formed first and had longer evolutionary histories. Planets closer to the Sun were more primitive. We can trace this back to the "nebular hypothesis" of solar system origins formulated in 1796 by the French astronomer Pierre-Simon de Laplace. According to Laplace, the planets were cast off sequentially as rings of gas from a shrinking, spinning cloud surrounding the Sun. Each gas ring eventually condensed into a new planet, with the innermost planets forming last. So, the theory suggests, Mars has been here much longer than Earth, and Earth much longer than Venus. Although this hypothesis has been replaced with new ideas about planet formation in greater accord with space age data (see Chapter 4), its vestiges greatly influenced science and science fiction until quite recently. Mars was often portrayed as past its prime, like a future vision of a dying Earth, with ancient civilizations nobly struggling against global change. Lowell explained the "canals" as a desperate attempt to survive the desertification of a dying planet by bringing water from the polar caps with a massive engineering project. Venus was consistently portrayed as "Carboniferous Park," a living replica of Earth's past. Venus, so near and yet always hidden by clouds, so Earth-like in every known aspect, helped to perpetuate for centuries a wide-spread belief in a universe abounding with populated worlds.

LOST OCEANS OF THE GOLDEN AGE

The scientifically supported doctrine of a plurality of worlds was fertile ground for metaphor and social commentary. From Voltaire's tales of planet-hopping characters who served as foils for his cynical view of human nature, to the visions of other worlds in early-nineteenth century Romantic poetry, philosophers, poets, and fiction writers seized on the vision of a cosmos filled with a multitude of inhabited planets, all acting out illuminating variations of the human tragicomedy.

The birth of modern science fiction probably occurred sometime in the mid-nineteenth century.* This literature was inspired but not constrained by the scientific revolution. Travel to other worlds and extraterrestrial life became two of its great themes. Science fiction writers have always sought to shape their extrapolations of the future to be consistent with the latest scientific ideas of the day. It is instructive, and fun, to trace the evolving past of the future, and of other worlds, as mirrors of the evolution of scientific ideas and social mores and attitudes. The so-called "golden age" of American science fiction in the late 1930s through the 1950s produced numerous tales of future generations of humans (men, mostly) exploring various manifestations of mysterious, cloud-veiled Venus. Most common among these visions are variations on the theme (exemplified by the passage from Bradbury that begins this chapter) of Venus as Earth's warm waterlogged twin, often crawling with more primitive versions of Earth life. Portrayals of this soggy pseudo-Earth range from idyllic Edens to frightening and hostile jungles.

In Isaac Asimov's 1954 novel *Lucky Starr and the Oceans of Venus*, the hero is a member of the powerful and secretive Council of Science.

---

*Mary Shelley's *Frankenstein* (1818) and H. G. Wells's *The Time Machine* (1895) are often named as the seminal works. Edgar Allan Poe and Jules Verne also get honorable mention as innovators of the form.

The council, which seems to be a combination of scientific society and secret police, defends civilization from the forces of darkness, and spreads science, reason, and justice throughout the galaxy. Mysterious trouble has been reported in the Venusian colonies, and Lucky is dispatched to clear things up. In a lecture to his sidekick as they descend toward Venus, Lucky explains:

> Until the first explorers landed on Venus, all mankind ever saw of the planet was the outer surface of these clouds. They had weird notions about the planet then. They weren't even sure about the composition of Venus' atmosphere. They knew it had carbon dioxide, but until the late 1900s astronomers thought Venus had no water. When ships began to land, mankind found that wasn't so.

The Venusian colonists live under domed cities built in the shallow parts of the oceans. The characters are all men, except for the ornamental wife of one of the principals who is briefly described as young and pretty, lives to decorate their apartment, and complains that she has to miss all the action. Something has gone horribly wrong on Venus, and Lucky and the other men of Science must navigate its deep, planetwide oceans, successfully battling Venusian frogs with telepathic power. In one terrifying scene Lucky and his cohorts encounter a giant killer jellyfish. Although the Venusian frogs get the upper hand and even briefly take over Lucky's mind, rest assured that Science eventually triumphs, restoring order and making the solar system once again safe for peace, justice, and the American way.

Another favorite is the camp 1958 film *Queen of Outer Space*, which stars Zsa Zsa Gabor as a brilliant Venusian scientist. This time the Venusian scientists are all women, although they are always dressed in evening gowns and pay far more attention to their appearance than most scientists I know. A group of men from Earth crash-lands on Venus and, resourcefully, they seduce Zsa Zsa and her band of renegade scientists, who then save them from their evil (female) dictator when they realize that "vimmen cannot live wizzout men." It is interesting to see how Venus's role as the only female planet in Western mythology is sometimes echoed in the gender roles assigned to Venusians in American pop culture (see Figure 2.7).

2.7 Zsa Zsa Gabor as a Venusian scientist in the 1958 film Queen of Outer Space.

Perhaps it is not a coincidence that the "golden age" of science fiction ended with the advent of the "golden age" of planetary exploration in the 1960s. As we began to explore the solar system, starting with Venus, the solid citadel of clouds that had protected all these fantasy worlds from intrusion began to crack under the persistent attack of science. And early results suggested that all was not comfortably warm and wet in the jungle.

3

in the time of spacecraft: descent into hell

You are not expected to complete the work in your lifetime. Nor
must you refuse to do your unique part.
—TALMUD

After such knowledge, what forgiveness?
—T.S. ELIOT, *Gerontion*

COMPARATIVE PLANETOLOGY

Aliens have invaded our solar system. Beginning in 1961, a fleet of
giant metal insects left the third planet and wandered inquisitively
throughout this planetary system. We made them and sent them on
their way.

Astronomy has been called the oldest science and Medicine the
youngest, but younger still is Astronomy's progeny, the fledgling field
of Planetary Science. The planets have been the subject of careful
observation and myth for millennia, and the subject of telescopic studies for centuries. Yet only in the last few decades have we been able to

send spacecraft, small mechanical extensions of our own senses, sailing through the solar system. As reports stream homeward at the speed of light, visions of a plurality of worlds, of a universe full of planets identical to our own and populated with beings just like us, are being rapidly displaced by new images of real worlds stranger and more diverse than we could have imagined. This newfound ensemble forms the subject matter of a new field (planetary science), and a new approach (comparative planetology) which have only come into their own in the age of spacecraft.

Historically, planets were part of the turf of astronomy and, as long as telescopes were the primary tools of discovery, they fit there comfortably enough, a roving subset of all the points of light in the sky. But as spacecraft began returning detailed pictures of alien landscapes and other troves of information, the techniques and concerns of astronomers were no longer adequate for the task of making sense of the planets. What was needed was the expertise of those who had thought in detail about the one planet that had thus far generated any detailed science—Earth. Whom do you consult when trying to understand volcanic flows on the Moon, Mars, or Venus? Why a volcanologist, of course. And if you want to understand the weather on Mars or the greenhouse effect of Venus, you don't start from scratch. You take an existing Earth theory or model and modify it, changing physical parameters to match those of the world you are studying.

In the 1960s, planetary science started employing the techniques of, and seducing a few scientists from, the earth sciences (geology, geophysics, geochemistry, meteorology, and others). Both groups of scientists benefited enormously from the cross talk. Planetary explorers got preexisting frameworks and techniques for studying what would otherwise be a bewildering abundance of new information and phenomena. They inherited the necessary expertise to take on the problems of planetary origins and evolution. Earth scientists got to judge how good their models were by stretching them to encompass whole new worlds, generating tremendous insights about Earth, and—in those cases where the models bent but did not break—new confidence in the results. We now have something to "compare and contrast," as the test questions require. And we can see our planet more clearly, mirrored in the freshly unmasked faces of our dancing partners as together we circle 'round the Sun. If we can successfully explain the differences among

Earth's climate, torrid Venus, and frigid Mars, then we have more reason to trust our predictions of climate change here, predictions loaded with significance for humans and the other species with which we share this world. Thus we all stand to gain in very practical ways from this exchange—not to mention the historical, spiritual quest to know our universe, and thus ourselves, better.

The danger in having so much of planetary science grow out of the field of earth science is that we may be overly predisposed toward "geocentric" interpretations: we sometimes assume Earth to be the standard against which other planets are measured, rather than simply one among many possible outcomes of planetary evolution. How could we possibly avoid some such bias?* Yet the growth of planetary science has also produced an enlarged, less provincial perspective on Earth.

First-generation planetary scientists, bringing with them the expertise, habits, and languages of many fields, have created an interesting cultural amalgam, with lots of communication and no small amount of miscommunication between people used to thinking differently from one another.

Most of the first generation are still around, actively pursuing research and training the next generation of explorers. My own experience illustrates how new this field is. When I started graduate school at the Planetary Sciences Department of the University of Arizona, where I worked on my doctorate from 1982 to 1989, none of our professors had degrees in planetary science, because when they were in grad school there was no such thing. Their doctorates were in geology, geochemistry, physics, meteorology, and astronomy. Now the faculty of planetary sciences departments have been partially infiltrated by people, like myself, with degrees in planetary science itself. I guess that makes us the second generation.

THE CATHEDRALS OF JANUS

If seeds in the black Earth can turn into such beautiful roses what might not the heart of man become in its long journey towards the stars?
—G. K. CHESTERTON

---

*However, the early earth science people who went into planetary science were those who were least likely to fall victim to "terrestrial chauvinism."

Planetary missions are cathedrals of our time. They will outlive by centuries all who participated in their making, and will be remembered for as long as there are humans, or humans are remembered. For many of us who grew up during the 1960s, 1970s, and 1980s, the solar system has been a magically expanding place, with horizons that receded as we grew older, and vague mysterious worlds that crystallized and clarified before our very eyes as we traveled through life. Many of us can play "I remember the moment when . . ." with the space program, as if experiencing the first human steps beyond Earth were a normal part of growing up.

One of my earliest vivid memories is from the fourth grade when my parents let me stay up way past my bedtime to watch a televised image of Neil Armstrong climbing down a ladder to make "one small step" onto the dusty surface of the Moon. In the summer of 1976 my cousin and I, both sixteen, were living in a tent in the woods outside of Stowe, Vermont. On the day that *Viking Lander I* was to set down on Mars we drove frantically into town just in time to see the first pictures of the rocky Martian surface emerge, slowly, in thin vertical strips, across the television screen of the local country store. The following year I graduated from high school and *Voyager* was launched toward the outer solar system. *Venera 12* landed on the surface of Venus on my eighteenth birthday, as *Pioneer Venus* orbited overhead, bouncing down radar echoes to make the first topographic map of the planet. By 1979, when *Voyager* reached Jupiter, I was an undergraduate research assistant at the Jet Propulsion Laboratory in Pasadena, where the frenzy and exhilaration of the *Voyager* encounters took place. As *Voyager* rapidly approached Jupiter or one of its exotic, icy moons, the latest image would appear, bigger and clearer than the last one, in the ubiquitous television monitors at JPL. It felt as though we were on the ship itself, looking at on-screen images of rapidly approaching never-before-seen worlds. And in a sense we were on board. Our expanding consciousness rode through the solar system with the *Voyagers*.

For the later *Voyager* encounters with Uranus in 1986 and Neptune in 1989, I was back at JPL as a graduate researcher.* Not everyone gets to see historic discoveries of this proportion unfold during their

---

*I missed Saturn in 1980. I don't remember why. I must have been hitchhiking to a Grateful Dead concert or doing something of equal importance.

formative years but, thanks to the magic of the mighty TV, millions got to watch it happen. Our worldview has changed and will continue to do so as the reality of what we have seen sinks in. The Copernican revolution is real. We've got the pictures to prove it—pictures of strange, beautiful places where the same physical laws that made our world, and us, have unfolded in wonderfully different ways to produce peculiar landscapes and atmospheres that strain our imagination, and our science. We are all richer for this, and I feel grateful to those who have had the vision, ingenuity, and perseverance to make it happen. It is undeniably a fantastic time to be alive, at least if you are born among the haves. And yet . . .

And yet there is a dark side to all this. We live in an age of maddening ironies, a time when technological marvels seem unlimited and yet much of our species is starving or held hostage to warmongers, a time when our understanding of planets has grown immeasurably but our ability to sustainably manage our home planet has never seemed less certain. For me these ironies are encapsulated with intense poignancy in the different purposes for which space technology has been put to use. Our fantastic voyages to the Moon and planets have been brought to us, at least in part, courtesy of Nazi rocketeers and cold-war paranoids. The same nation that produced *Apollo*, *Mariners*, *Vikings*, *Pioneers*, *Voyagers*, *Galileo*, and *Magellan* also brought us the Vietnam War and has been the largest exporter of weapons to developing nations. The same folks that brought us the Mir Space Station and the *Venera* and *Vega* missions to Venus and comet Halley also built the Berlin Wall, invaded Afghanistan and came in a close second at providing arms to places where food is desperately needed. This is no coincidence. Space exploration and war are inextricably linked. Many of the technologies needed are the same, and the same corporate contractors have been leaders in both arenas.

The Apollo 11 astronauts left a plaque on the Moon stating that "We came in peace, for all mankind." Each part of this statement is at best a half-truth.* Although the ostensible motivations for the man-on-the-moon program, and certainly its most lasting benefits, were peaceful scientific research and exploration, it never would have happened if we weren't trying to remind our Cold War adversaries that we

---

*That doesn't add linearly to make one truth, does it?

had the technical prowess to bomb them back into the stone age. And although the moon landing may have been "for all mankind" (at the time it was the most widely watched event ever on live TV) it certainly wasn't *by* all humankind. It was mostly done by white men, unless you count the people sweeping the floor after hours in Houston.*

I have met many people who feel alienated from the space program for these reasons, the military connection in particular. When I worked as a post-doctoral researcher at NASA's Ames Research Center I got used to a certain kind of reaction from people who, when learning of my NASA affiliation, would suddenly start to act as if they had just met one of Hitler's deputies. It saddens me that these people cannot appreciate the wonders that our explorations beyond Earth have revealed to us. I certainly see where they are coming from. The Apollo Moon rocket was built by many of the same engineers who built V-2's to bomb London during World War II. But NASA and the space program are not monolithic entities. Many people are motivated by curiosity and environmental activism. In fact, Apollo, for all its militaristic posturing, also served as a major catalyst for environmental awareness. Photographs of Earth from distant space forever changed the self-image of Earth's inhabitants, helping to awaken our latent planetary identity (see Figure 3.1). The different uses of space technology are an expression of the two-faced, Janus-like nature of humanity.† We have no choice but to live with these contradictions, and try to be a voice for the positive side of the force.‡

If the planetary missions are our cathedrals, are there lessons to be learned within which can help us to guide ourselves, to live wisely? I believe there are. Although we sailed to the other worlds on winds of war, we returned with essential knowledge of the workings of planets. This knowledge and perspective may ultimately help us to live in peace on, and with, our own world.

---

*The number of women working in the space program has increased considerably. There are still not a lot of people of color involved.

†Janus is a two-faced Roman God often used to symbolize dualities. He is the porter of Heaven and has one face for the rising sun and one for the sunset. Perfect for Venus watching.

‡Comedian Mort Sahl expressed this irony with his sarcastic title for an autobiography of former Nazi rocket scientist, and Apollo architect, Werner Von Braun: "We aim for the stars, but sometimes we hit London."

3.1 Earthrise.

RADAR AND ROCKETS

It is hardly new for the builders of cathedrals to be also the masters of war. Throughout our history, peaceful, curiosity-driven scientific research has often ridden piggyback on seemingly more urgent, well-funded military research. Galileo's telescopic observations, from which sprang the 350-year fantasy of a twin Earth beneath the clouds of Venus, received support from Venetian* senators impressed by the telescope's military potential. The discoveries that finally woke us from this convoluted dream came from technological advances that germinated during World War II and blossomed, energized by cold-war competition, in the 50s and 60s. Chief among these inventions were radar and rockets.

Radar, the use of reflected radio waves to detect distant objects, was developed rapidly on both sides during World War II as a way to detect planes, ships, and rockets made invisible by darkness, distance, or clouds. After the war the military secrecy surrounding the technology was lifted, and scientists began to explore its research potential. Radar astronomy was born in 1946 when U.S. Army scientists bounced radar waves off the Moon and detected the return signal.

Radar signals were first successfully reflected off the surface of Venus in 1958. In 1962, radar measurements allowed us to resolve the centuries-long argument over the true rotation rate, which was described in Chapter 2. It turned out, once again, that everyone was wrong. Venus rotates extremely slowly, so that one day there is equal to 117 Earth days. And it turns in a backward, or "retrograde," direction compared with most other planets. If you were standing on the surface of Venus, you would sense the Sun rising in the west and setting in the east 59 (Earth) days later. I say "sense," not "see," because it is always overcast on Venus, everywhere, every day. (Weather forecasters on Venus would have a much better track record than they do here.) But dawn brings a brightening in the west that moves ever so slowly across the sky as the two-month-long day unfolds. Every Venusian year the Sun rises and sets twice.

---

*That's *Venetian* as in "from Venice." The Venusian Senators are a baseball team from Venus who were not even born in Galileo's day.

A few years earlier, in 1956, passive radio observations* had allowed a startling discovery that led to the first serious challenge to the fantasies of a warm, wet Venus. Mysteriously, the planet was emitting very large amounts of microwave radiation. Further observations in the early 1960s confirmed its brightness when viewed with microwave eyes. How could this be? If Venus is Earth's twin, why is she leaking radiation like a broken microwave oven? Some researchers inferred that this was heat radiation from the surface. If the microwaves were the result of a glowing, hot surface, this would imply a surface temperature of over 600 degrees Fahrenheit. But the clouds were still thought to be made of water vapor, and water clouds, along with oceans and life, require a relatively cool surface. Many people resisted the hot surface explanation, and several alternative explanations were proposed for the microwaves. Could the oceans of Venus still be saved by finding another source for the microwaves, perhaps emissions from a hot upper atmosphere, or radio noise from massive lightning storms?

As long as Venus remained an object of distant observation in our sky, there was no way to be sure, and science fiction writers were free to populate Venus with ocean-dwelling beasts and evil dictators (news of the problematical microwaves was first published the same year that Zsa Zsa was thrilling audiences with her Venusian antics). We had to go there to demand some answers. This is where the rockets enter the story.

The rocket designers who made V-2's for Hitler to "sometimes hit" London, employing slave labor in their construction, were eagerly recruited after the war by both the United States and the Soviet Union. Each side felt it important to demonstrate to the other that it could build the powerful rockets and reliable guidance and atmospheric entry systems necessary to deliver nuclear warheads. Fortunately for the scientists who were ready to begin exploring the solar system, successfully launching a spacecraft to another planet is an excellent way to demonstrate such prowess. The cold-war "space race" was on, and science was along for the ride, onward to the planets.

---

*Meaning that no signal is sent, and so an object must be emitting radio waves to be observed.

LUCIFER FALLS AGAIN

Venus was the target for the first spacecraft sent from Earth to another planet, *Mariner 2*, launched on August 27, 1962. One of the primary technical concerns was that we maintain two-way communication with the spacecraft, and keep enough control over its trajectory to keep it from crashing into the planet and possibly contaminating Venus with terrestrial microorganisms. For all we knew, Earth germs might like it on Venus and compete unfairly with the Venusians.

In addition to studying its primary target, Venus, *Mariner 2* measured for the first time the properties of the unknown, unexplored environment of interplanetary space, including magnetic fields, the solar wind, and cosmic dust. En route to Venus, *Mariner 2* (see Figure 3.2) was hit by exactly one speck of dust large enough to detect, showing that the inner solar system was relatively free of a micrometeorite threat.

On board was an instrument called a microwave radiometer. This was designed to observe how the mysterious microwaves varied in intensity across the face of Venus and to resolve the loaded question of whether they came from a sizzling hot surface or some strange atmospheric phenomenon. As *Mariner 2* approached Venus, scientists were of two minds, with two clearly different predictions for the pattern that this radiation would make, depending on where it was coming from. If a hot surface was the source, as the spacecraft flew by it should record a peak coming from the center of the planet, where radiation from the surface would be least blocked by the atmosphere. In this case there would be less of a microwave glow observed at the edges, or limbs, of the planet, where it is filtered through a greater thickness of atmosphere. This phenomenon is known as limb darkening. If the radiation was being produced *in* the atmosphere (which would allow for a cool, wet surface), the opposite effect was predicted: a limb brightening of the microwaves, with the most intense radiation at the edges.

Allow me a nerdly illustration: Consider your view from above the head of a man with a crew cut. You see more scalp, peeking through his short hair, at the center of his head. The scalp is "limb darkened" because in the center you have a more direct view of it, whereas at the sides, due to the shallower angle of your view, there is more chance for hairy obscuration. However, if it's hair you are looking for, the opposite

3.2 The **Mariner 2** *spacecraft. (NASA)*

is the case. The hair is "limb brightened," meaning you see more of it at the edges and less at the center. *Mariner 2*'s microwave radiometer was designed to see whether the great microwave brightness was coming from the skin or the hair of Venus (see Figure 3.3).

Microwave limb darkening means a hot, dry surface. Limb brightening means that oceans and all they imply are still possible. It is rare in planetary science that a single observation is made with such a clear choice of predictions that can instantly resolve between two radically

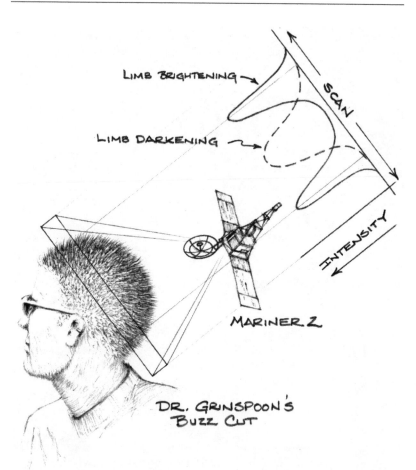

LIMB BRIGHTENING

LIMB DARKENING

SCAN

INTENSITY

MARINER 2

DR. GRINSPOON'S
BUZZ CUT

3.3 Limb brightening vs. limb darkening. *The scalp is "limb darkened"; you see more of it in the center. The hair is "limb brightened"; you see more of it at the edges.* **Mariner 2** *showed that the microwave radiation of Venus is limb darkened, and so must be coming from a hot surface. (Carter Emmart)*

different views of a planet's nature. Much more often the gains are incremental and the implications of new observations take years to hash out. But on December 14, 1962, *Mariner 2* flew past Venus at a distance of twenty-one thousand miles, and the microwave radiometer worked perfectly. The results showed a clear pattern of limb darkening. (More microwave radiation was coming from the center of the planet than from the edges.) This meant that the source of the intense microwaves was at the surface, proving that it must be scorching hot down there.

In fact, the surface temperature is 735 degrees Kelvin, or nearly 900 degrees Fahrenheit. You could fry an egg on the sidewalk, but you'd have to do it quickly, before the sidewalk melted. It's hot enough there to melt many spacecraft materials and fry electronics.* Any water on the scalding surface would boil instantly. So much for our "twin" planet's global oceans. Alas, no giant jellyfish, tree ferns, or dinosaurs. No telepathic frogs, at least not carbon-based ones, and things suddenly looked a great deal more dubious for those "millions of intelligent creatures" vibrating the air with their musical tongues. Life, or at least "life as we know it," based on carbon molecules dissolved in water, is clearly impossible at the surface of Venus. If we were put there without a very good climate-controlled suit, we would quickly and quite literally be dead meat. The water in each of our cells would instantly boil off, and our proteins would quickly shred into tiny fragments that would react violently with the surrounding gases. Venus is no place to raise the kids. In fact, it's hot as hell.

Another important tool for remotely probing the composition of our planetary neighbors is *spectroscopy*, which starts with the common-sense notion that different substances are different colors and extends into realms beyond our senses. The radiation† we call visible is a tiny portion of the much larger electromagnetic spectrum, which extends beyond the violet limit of our eyes out to ultraviolet light, and X rays, and beyond the red to infrared, microwaves, and radio. The Sun shines mostly in visible light. This is no coincidence. We have evolved orbiting a moderately hot star, and our eyes have adapted to make good use of the radiation from a star at this temperature. Just beyond the red edge of the visible lies the infrared, light with wavelength too long and energy too low for our eyes to see (see Figure 3.4).

Whereas atmospheric gases tend to be conveniently transparent in the visible range (allowing us to see the Sun and the stars, and everything else by reflected sunlight), many gases absorb infrared light. Each has a distinct pattern of absorption at specific wavelengths—a unique molecular fingerprint. With infrared spectroscopy you analyze the infrared color spectrum in great detail, note which wavelengths are

---

*Early Soviet probes carried aluminum busts of Lenin to the surface. The surface temperature is higher than the melting point of aluminum, so these must have quickly become small revolutionary puddles.

†"Radiation" and "light" are the same thing.

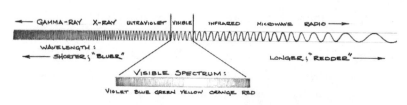

3.4 The Electromagnetic Spectrum: *Visible light is a small part of a continuous spectrum that extends to wavelengths longer and shorter than those our eyes can see. (Carter Emmart)*

being absorbed, and thus decipher the chemical composition of distant atmospheres. But first you must carry your telescope above as much of Earth's atmosphere as possible. In particular, if you want to observe the absorption signature of water on Venus, you have to get up to a height where it is so cold that most of Earth's obscuring water vapor has frozen out. Thus a high, cold mountaintop is a good choice. The first spectroscopic observations of Venus, made in 1932, showed no signs of water, but the spectral signature of carbon dioxide was clear. Increasingly refined observations throughout the 40s and 50s also showed evidence only for CO_2.

So where was the water on this cloud-covered world? In the early 1960s observers started to carry telescopes equipped with steadily improving spectrometers up in balloons and aircraft that provided colder and clearer vantage points than any mountaintop. At last these high-tech dowsers achieved their goal: they found the infrared fingerprints of water vapor in the air above the clouds of Venus. They were also able to measure the amount of water, but these results were strange. For one thing, several of the different water measurements conflicted widely. For another, they all suggested that it was far too dry for air sitting on top of water clouds. Typical measured values suggested only one or two parts per million of water vapor above the clouds. If this were right, it seemed to rule out clouds made of water or ice.

Mariner 2 had shown that the surface of Venus is hotter than an oven. Spectroscopic observations made during the same time period, and later direct measurements by American* and Soviet entry probes, revealed that the atmosphere of Venus is also bone dry, casting further

---

*I am going to refer to the space exploration missions of the United States as "American" missions. I recognize the limitations of this term, but it flows so much better than "United Statesian."

doubt on the widespread notion that the clouds of Venus were Earth-style water clouds. Remember, these clouds were all that we had ever actually seen of Venus. They give Venus her myth-making brightness in our sky. They enshroud the entire planet, and they were largely responsible for the sister planet fantasy (clouds = water = oceans = life = civilization [?]), and for its persistence (because you can't easily dismiss what you can never see). But with such a minuscule amount of water in the Venusian air, the clouds couldn't be made of water droplets. So what were they?

The quest to identify the mysterious cloud-forming substance took another decade after *Mariner 2*. Numerous observations and rampant speculation paraded a menagerie of exotic candidates through the scientific literature before we found the answer. These included formaldehyde, plastic polymers, rock dust, oil droplets or other organic compounds, and salt. The answer, when we finally found it in 1973 from a combination of spectroscopy and polarimetry (studying how the cloud droplets polarize light), was at least as strange as any of these: the clouds are made of highly concentrated sulfuric acid, commonly known as battery acid.

Thus we found our "sister planet" to be chemically alien, as well as hot and dry to quite unearthly extremes. With these revelations, the twin-sister imagery quickly disappeared, and the notion that "Venus is hell" took hold. Once again, Lucifer had fallen.

This was a time of furiously paced space exploration, aggressive and optimistic. Less than a decade after *Mariner 2*, in 1971, *Mariner 9* revealed ancient water-carved channels on Mars, and Mars became our next best hope for finding Earth-style life, or at least fossils of past life, elsewhere in our solar system. If we could no longer imagine teeming oceans obscured by the clouds of Venus, we could at least imagine oceans in a more illustrious past on Mars, obscured instead by time, with the ancient, dusty, dried-up river valleys revealed by *Mariner* and *Viking* perhaps being all that remains of a mighty Martian hydrological (and biological?) system. The three-hundred-year vision of Venus as Earth's twin has faded to a quaint relic of Victorian-era romanticism.

With the clarity of hindsight, it is easy to discount the twin-sister vision of Venus as so much wishful thinking based on rather flimsy evidence, although this is rather unkind to some of the best observers and thinkers of the last three hundred years. Our telescopes gave us just enough information to fuel this fantasy and not enough to dispel it. Maybe we became a bit too comfortable with, and comforted by, this vision of another Earth, perhaps an Eden, so nearby. It was a fantasy created by our own desires and supported by science. We put Venus on a pedestal and set ourselves up to be disappointed, in a way that is not quite fair to Venus or ourselves. When our fly-bys, orbiters, and landers started to pry, poke, and peek beneath the clouds, it was against this preconceived notion that Venus was judged. Once it became clear that the environment on Venus was not as familiar and friendly (to us) as had been imagined, the language and imagery used to describe our neighbor world transformed radically and rapidly.

The new truth was "the grim truth," and Venus, as if she had betrayed us by not living up to our expectations, was judged harshly. "Venus is hell" became a common expression, and words like "harsh," "hostile," "inhospitable," "errant twin," "poisonous," "catastrophe," "noxious," and "tortured" filled the pages of books and articles describing our closest sibling. Nearly every book, chapter, or article written about Venus since the 1960s contains some version of the statement: "The planet named for the Goddess of Love turned out to have a closer resemblance to Dante's Hell."

In Arthur C. Clarke's visionary 1953 science fiction novel *Childhood's End*, an advanced race of aliens visits Earth. But they must conceal their physical appearance from humans because of their uncanny resemblance to mythic images of the devil, complete with horns and pointed tails. Venus may suffer similar guilt by association with biblical and literary images of hell, at least as far as being deathly hot, dry, and full of sulfur vapors.* But after this point has been made, isn't it awfully Earth chauvinist to keep hammering the "Venus is hell" theme? That life based on organic molecules would not fare well there is true: sulfuric acid and 900 degree heat are not healthy for carbon-based children and other organic living things. But neither is any other environment we have discovered elsewhere in the solar system, or anywhere else. It is natural to be extremely attached to the environmental conditions to which we've become accustomed after living, and evolving, here for 4 billion years. But if we are to harshly condemn every environment

---

*The religious association of sulfur vapors with hell probably comes from identifying volcanic vents as gateways to the underworld.

where we cannot live comfortably, the universe will seem a lonely, sterile, and hostile place. And, especially when we see the great complexity, beauty, and familiarity of Venus as revealed later by *Magellan* and other more recent developments, it seems that "Venus is hell" may have been in part an overreaction to the disillusionment of learning that "Earth's twin" was a naive dream.*

Science fiction writers were quick to update their Venus imagery. Larry Niven's *Becalmed in Hell*, published in 1965, depicts a troubled cyborg ship stalled in the sweltering Venusian depths, its terrified pilots expecting to be cooked. Niven assumed that at the surface it's "as black as it ever gets in the solar system." As we will see, conditions down there are not quite that dim.[†]

STORMING THE HEAVENS OF HELL

Your mysterious mountains I wish to see closer. May I land my kinky machine?
—JIMI HENDRIX, "Third Stone from the Sun"

After *Mariner 2* the people of Earth, or at least the two nations caught up in the cold war, began sending probes to Venus in steadily increasing numbers and sophistication. We often refer to this period wistfully as the golden age of planetary exploration. It was a time of relative affluence and optimism in the United States. A popular young president had declared that we would put men on the Moon by decade's end, leading to an ambitious and lightning-fast, by today's standards, program of rocket design and testing. And, of course, there was that cold-war rivalry giving both sides a sense of urgency in maintaining a technological edge and achieving "firsts" throughout the solar system. For all of these reasons it was much easier then than it is now to get a new planetary mission approved, funded, and flown. Each new space-

---

*Venus may seem like hell to us, but of course we are deeply, profoundly biased in this assessment, perhaps more than we know or can know. Alien tastes and aesthetics may be very different. Who knows, maybe they'd like the smell of microwave popcorn and the sound of fingernails scraping on a blackboard. And the inspiring music of Bach or Bob Marley might make them want to puke, if that's something they do.

[†]Niven's story has people going to Venus in 1975, reflecting the optimism of extrapolations during the *Apollo* era. The ship in this story, suffering from psychological problems, is a predecessor to the paranoid computer in Clarke's *2001: A Space Odyssey*.

craft was larger and carried more scientific instruments than the previous one—nearly the opposite of recent trends.*

Thirteen hundred years ago Mayan astronomer-priests eagerly anticipated and accurately predicted the short disappearance interval when Venus would vanish from his evening apparition and reemerge, eight days later on average, from the underworld as morning star. They knew that every nineteen months this drama, with subtle variations that they charted and cherished, would repeat. Today we see this time of disappearance and reemergence as "inferior conjunction," the time when Venus in her orbit passes between us and the Sun. Now this configuration is eagerly anticipated by spacecraft engineers and planetary scientists, because it represents a new kind of opportunity in the continuing quest for human connection with the lights in the sky. For the Maya these events were times of spiritual renewal, sacrifice, and battle; for us, they are launch windows. On these occasions, at nineteen-month intervals, the two planets swing closest, and this means that a spacecraft launched near that time requires the least energy (least fuel) to reach Venus. It also makes for shorter travel times, which means less chance of something going wrong along the way, and a shorter radio communication distance from spacecraft to Earth, allowing for maximum data retrieval.

Table 3.1 shows the dates of Venus inferior conjunctions from the time of *Mariner 2* in 1962 to the launch of *Magellan* in 1989, along with the spacecraft we launched at each of these opportunities, and the date of launch and Venus encounter for each. You can see that inferior conjunction always occurs between launch and encounter. This is because we achieve the minimum energy trip if we launch a spacecraft from an orbital position about 60 degrees ahead of Venus as she closes on us in the inside lane. The spacecraft is always en route on the actual date of inferior conjunction, so we launch toward "evening star" and arrive at "morning star" (see Table 3.1 and Figure 3.5).

This table indicates some interesting differences between the United States and the Soviet Union in space exploration style. Notice that, beginning with the unsuccessful 1961 mission *Venera 1* (which would have been the first mission to Venus had it made it there), the Soviets took advantage of at least every other launch window from the 60s through the 80s. They had many failures, particularly early on, but

---

*But our instruments keep getting better.

3.1 VENUS INFERIOR CONJUNCTION AND SPACECRAFT LAUNCHES (1961–1990)

INFERIOR CONJUNCTION	SPACECRAFT	LAUNCH DATE	ARRIVAL AT VENUS
4/11/61	Venera 1	2/12/61	5/19/61
11/12/62	Mariner 2	8/27/62	12/14/62
6/19/64			
1/26/66	Venera 2	11/12/65	2/27/66
	Venera 3	11/16/65	3/1/66
8/29/67	Venera 4	6/12/67	10/18/67
	Mariner 5	6/14/67	10/19/67
4/8/69	Venera 5	1/5/69	5/16/69
	Venera 6	1/10/69	5/17/69
11/10/70	Venera 7	8/17/70	12/15/70
6/17/72	Venera 8	3/27/72	7/22/72
1/23/74	Mariner 10	10/3/73	2/5/74
8/27/75	Venera 9	6/8/75	10/22/75
	Venera 10	6/14/75	10/25/75
4/6/77			
11/7/78	Pioneer Venus Orbiter	5/20/78	12/4/78
	Pioneer Bus and Probes	8/8/78	12/9/78
	Venera 11	9/9/78	12/25/78
	Venera 12	9/14/78	12/21/78
6/15/80			
1/21/82	Venera 13	10/13/81	3/1/82
	Venera 14	11/4/81	3/5/82
8/25/83	Venera 15	6/2/83	10/10/83
	Venera 16	6/7/83	10/14/83
4/3/85	Vega-1	12/15/84	6/11/85
	Vega-2	12/21/84	6/16/85
11/5/86			
6/13/88			
1/19/90	Magellan	5/4/89	8/10/90

at nearly every opportunity they would throw whatever they had ready at Venus. The difference between one spacecraft and the next, in the Soviet *Venera* series, tended to be incremental, with each year's model closely resembling the last one but incorporating whatever improvements in instrumentation had become available since the last launch.

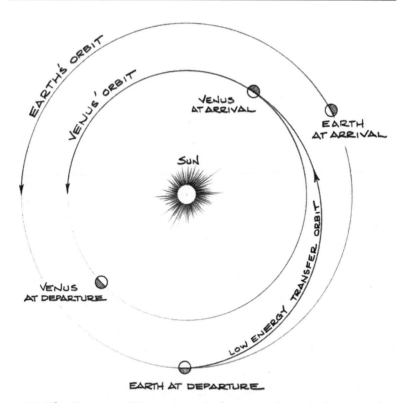

3.5 The Route to Venus: *A typical spacecraft path from Earth, designed for minimum energy use during transit. (Carter Emmart)*

The American launches were fewer and farther between but more sophisticated technologically, and they had a higher success rate.

These differences may reflect the different political systems of the two space-faring superpowers. In the Soviet Union, with its highly centralized government, a small number of people made decisions about space exploration. It was easier to get a mission approved there than in the United States, where every space mission must navigate countless NASA committees and then survive the perils of congressional funding cycles— a journey at least as treacherous as that through interplanetary space. Our missions are often threatened with extinction every year until launch, and some have been canceled and restarted several times. There is a mismatch between the timescales of conceiving, designing, and flying planetary missions and the pace of change in American politics. A mission that receives a green light from one presidential administration and Congress may get canned a few years later when the winds change in Washington.

The Soviet programs, once started, took on lives of their own. But, consequently, there were fewer incentives toward perfection and fewer checks on the system, no bidding between contractors eager to outperform each other, and no congressional oversight committees. So the American missions that did make it through to launch, although less frequent, were more carefully planned and more likely to succeed, and each tended to represent a major technological advance over the previous one. They had a "five-year plan"; we had tremendous "Yankee ingenuity," but little ability to stick to any plan. The two styles complemented each other, and throughout the 60s, 70s, and 80s our understanding of Venus slowly increased in depth and sophistication. But still the clouds did not clear, and major mysteries remained.

The early Soviet efforts first produced a string of failures that nonetheless provided valuable lessons in spacecraft design and navigation. *Venera 3* was designed to be the first spacecraft to land on another planet, measure surface conditions, and, of course, carry commemorative Soviet emblems to the surface of Venus. It actually reached Venus only to crash into the planet on March 1, 1966, returning no scientific data.* The Soviet's first successful mission, *Venera 4*, reached Venus in October 1967 and marked a major achievement: it was the first planetary probe to enter the atmosphere of another planet, do direct experiments, and radio home the results. *Venera 4* was protected from the intense heat of high-speed entry into the upper atmosphere of Venus by a heat shield developed through the testing of nuclear warhead reentry capsules for intercontinental ballistic missiles. During its brief (ninety-four-minute) descent by parachute, the spacecraft measured conditions in and below the clouds, confirming that the atmosphere of Venus is mostly carbon dioxide and recording increasing temperatures and pressures until it was crushed when it reached a level where the pressure is eighteen atmospheres (eighteen times Earth's atmospheric pressure), and the temperature is about 500 degrees Fahrenheit. Initially, Soviet scientists thought that *Venera 4* had reached the surface. Later they realized that the spacecraft had failed at an altitude of more than sixteen miles, meaning that the surface temperature and pressure must be considerably higher still. Evidently, future entry probes would

---

*So it was the first object from Earth to reach another planet, even if it didn't do anything useful there.

have to be built to withstand higher pressures, if we wanted to reach the surface. But how high? Would we ever find solid ground?

Conducted by the incessant five/eight orbital polyrhythm of Earth and Venus, the space-race contestants danced to a slow, nine-teen-month beat, with spacecraft launches concentrated into the regularly spaced launch opportunities that assumed this same rhythm. Nearly every time Venus swung near to Earth, a small instrumented probe (or several) would leap off Earth and try to hit Venus. Soviet and American spacecraft often literally raced each other to Venus and arrived within days of one another. Two days after the launch of *Venera 4*, the Americans launched *Mariner 5* from Cape Kennedy. As the Summer of Love turned to autumn back home, these two little machines from Earth raced toward Venus. *Venera 4* made its historic entry into the atmosphere of Venus on October 18, 1967, and the very next day *Mariner 5* flew past Venus at a distance of only twenty-one hundred miles with a battery of sophisticated new instruments trained on the atmosphere. Radio telescopes on Earth made precise measurements of its signal as it gradually disappeared behind the planet and reappeared on the other side. The *Mariner* data produced a much clearer profile of the atmosphere, revealing a surface temperature and pressure of about 980 degrees Fahrenheit and 100 atmospheres!

American and Soviet scientists shared information with each other quite freely and used each other's results to help interpret their own. The data from *Venera 4* and *Mariner 5*, when analyzed together, created a more complete picture of the atmosphere than either could have provided separately. Also, the Soviets, having had previous missions fail because of communication problems with their spacecraft, received permission to use the large radiotelescope at the British Jodrell Bank Observatory to monitor the signals from *Venera 4* as it plunged into the atmosphere of Venus. In these ways the space race acted as its own antidote to the aggressive, competitive forces that had spawned it. Planetary exploration does not recognize national borders. Planetary science is inherently a collective endeavor that at some level demands a planetary, not national, identity of those who practice it, as scientists from Earth try to understand our near neighbors.*

*Many astronauts have commented on the striking absence of political boundaries on Earth when seen from space. Something of the same perspective is demanded of those of us who are stuck down here looking up, sharing resources and attempting to unravel the story of the planets.

These first missions helped to confirm and refine the general picture of the Venus atmosphere that *Mariner 2* had sketched, a hot, dense envelope composed mostly of carbon dioxide. But many more intriguing questions were raised: What are the clouds made of and to what depth do they extend? Does the upper atmosphere really whip around the globe, rotating many times faster than the solid planet (as had been suggested by ground-based telescopic observations at ultraviolet wavelengths)? What other gases are in the atmosphere? Why doesn't there seem to be any water or oxygen? And what about the surface? What's it really like down there? Is it hot everywhere, or are there cool places at the poles or on mountaintops? Is the surface alive with active volcanoes, rugged young mountain ranges, and rapid erosion like Earth, or is it old, dead, and covered with impact craters like the Moon, Mercury, and the southern half of Mars? Was there ever water there?

Inquiring minds wanted to know, and they knew how to find out: U.S. scientists drew up ambitious plans for further Venus missions in the 70s, including multiple entry probes, orbiters, and a lander. However, the momentum got lost in a series of delays, budget cuts, and political shifts. Post-*Sputnik* hysteria, which had helped launch *Apollo* to the Moon and fueled early American planetary exploration, was subsiding. The bills from Vietnam and the cold-war military buildup were starting to come due, and the relative prosperity of the 60s was receding. The planners and designers of all subsequent American missions, up to the present day, have had to squander huge amounts of time and energy jumping through ever-shifting hoops of bureaucracy and fiscal readjustment.

Meanwhile, the Soviets launched successful entry probes at each of the next three launch opportunities, *Veneras 5* and *6* in 1969, *7* in 1970, and *8* in 1972. The latter two were the first Earth machines to land gently on the torrid surface of Venus. Designing and flying a spacecraft that can survive the descent through the ever-increasing heat and pressure to the surface of Venus and operate there even for a short while before succumbing is no small engineering challenge. The Soviets failed several times before succeeding. *Venera 7* made it all the way down only to fall over on impact with the ground. This made it hard to receive the signals because the antenna was pointed the wrong way. The operators on Earth heard *Venera* scream "I made it!" but did not retrieve any data from the surface.

They finally did it with *Venera 8*, which landed on July 22, 1972, and functioned on the ground for almost an hour, making the first direct measurements of the temperature, pressure, illumination level, and rock properties at the surface. As *Venera 8* descended toward the daytime surface, it recorded the decreasing amount of sunlight. A great deal of light was absorbed above an altitude of twenty-five miles, correctly attributed to clouds at these heights. Below about nineteen miles, the air was very clear. Measurements showed that about 2 to 3 percent of the Sun's light makes it down to illuminate the surface. This is really plenty of light to see by, equivalent to a deeply overcast day on Earth. It would be possible to take pictures at the surface, and the Soviets immediately adjusted their plans to include cameras on future landers. Wind speeds were found to decrease steadily with altitude, from over a hundred miles per hour in the clouds to just a few feet per second at the surface. The mixture of gases composing the air was found to be about 95 percent carbon dioxide, no more than 5 percent nitrogen and less than 0.4 percent oxygen. The air on Earth is made mostly of the latter two gases. The lack of oxygen on this nearby world is of particular interest to we who breathe it.

Although American exploration of Venus was stalled, at least temporarily, after *Mariner 5,* Venus scientists lucked out and got a free ride on *Mariner 10,* whose primary destination was the planet Mercury. *Mariner 10* needed to swing close by Venus to get a "gravity assist," slowing down by grabbing on to Venus's strong gravitational pull, which would drop it toward Mercury. Launched in October 1973, *Mariner 10* experienced a multitude of technical difficulties on the way to Venus, including flaky antennas, gyros, and power supplies. Several of these came perilously close to dooming the mission, but the American flight controllers, using their now legendary ability to apply on-the-fly ingenuity to fix spacecraft millions of miles out in space, kept the cranky craft running. It flew by Venus at a distance of thirty-six hundred miles in February 1974, and its cameras (designed to photograph Mercury) sent home the first close-up photographs of Venus from space.

At first these pictures revealed . . . nothing much. Seen from close by at visible wavelengths, the planetwide clouds appear basically bright and featureless, just as they do through a telescope. Photos of the limb (edge) of the planet did reveal several distinct layers of haze above the clouds in the planet's upper atmosphere, silhouetted against the blackness of space. But the real surprises came when *Mariner* took photographs through an ultraviolet filter; a dynamic and volatile new face of

Venus suddenly appeared. Observers on Earth had previously seen some vague features in ultraviolet images taken with telescopes, so the researchers expected some kind of markings. But what we actually saw far exceeded anyone's expectations: a complex swirl of high-contrast features, ranging from tiny, detailed splotches to huge planetwide streaks (see Figure 3.6). And the stuff moves around like crazy.

The identity of this material, so dark in the ultraviolet that it is responsible for nearly half the solar energy absorbed by Venus, is still not known, one of the great mysteries of Venus. Whatever this "unknown ultraviolet absorber" is, its motions allow us to trace the atmospheric currents in the cloud-top region of Venus. Like ink drops in water, or smoke in a wind tunnel,* the ultraviolet markings suddenly rendered the air currents visible. Meteorologists now had something to sink their teeth into, a new pattern of atmospheric circulation to model and compare with those seen on Earth, Mars, and Jupiter. Huge C- and sideways Y-shaped dark markings, symmetrical across the equator, were seen forming and dissolving. These all rushed by at a speed of nearly two hundred miles per hour, circling the whole planet in four days. This pattern of rapid planetwide rotation was dubbed the "superrotation." Finding its cause presented a major challenge for comparative planetology. Can our models of atmospheric circulation, developed on Earth but based on (hopefully) universal laws of physics, also be used to explain such an extremely different pattern of atmospheric motions on this nearby world?: another major mystery that persists to this day.

Other unusual atmospheric features seen in *Mariner 10*'s ultraviolet pictures of Venus, grist for the mill of the meteorological modelers, were bright hoods with spiral patterns extending from them over both poles, and a series of polygonal shapes that seemed to follow the point on the planet where the Sun was directly overhead, probably due to rising air from convection caused by solar heating. During its brief fly-by of Venus, *Mariner 10*'s other instruments gathered additional data that helped build a preliminary picture of the composition and thermal structure of the upper atmosphere and its interaction with the solar

*Or sneakers in the Pacific Ocean. In 1990 a shipping accident dumped sixty thousand athletic shoes into the North Pacific. Scientists opportunistically tracked and modeled the swarm of floating sneakers to study ocean circulation. A few teenagers were spotted in the Pacific Northwest wearing soggy, unmatched Nikes.

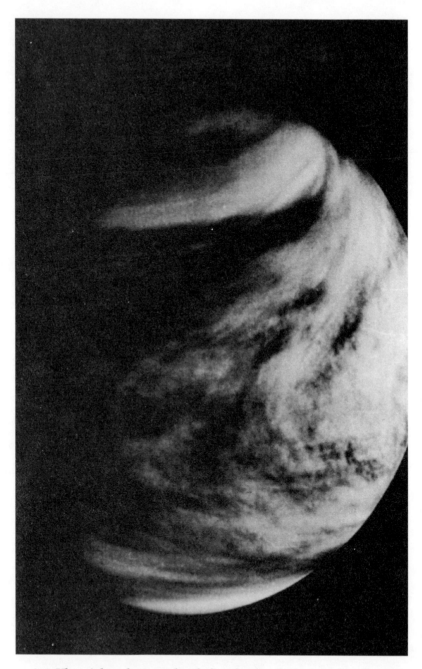

3.6 Ultraviolet photograph of the clouds of Venus taken by the Mariner 10 *spacecraft. (NASA)*

wind. Then, having received just the right gravitational kick from Venus, *Mariner 10* went on to Mercury, where it became the first and only spacecraft (so far) to visit the small innermost planet, successfully taking detailed pictures of half of its ancient, cracked, and pockmarked surface.

Unlike their American brethren who were tangled up in a draining game of "red light, green light" with the whims of successive administrations and Congresses, Soviet scientists and engineers did not have to wait to start planning their next Venus missions. At least while the cold war simmered, they had a mandate for a long-range plan to explore Venus. Soviet Mars exploration had produced nothing but failures. Once they started having a winning streak on Venus, the "reds" gave up on the "red planet" and concentrated their resources on Venus. This is odd, in a way, because the environment of Venus presents considerably greater technical challenges than that of Mars, at least for surface landers. It may have come down to luck, but once they set their sights on Venus they were relentless and, eventually, remarkably successful.

Emboldened by the success of *Venera 8*, they set about designing a new, more advanced generation of Venus craft. Each of these included a large, improved lander and an orbiter. The redesigned landers consisted of a spherical shell, eight feet in diameter, designed to withstand high pressures and containing interior layers of thermal insulation and coolant. They had a large disk-shaped drag plate, which, mounted on top to slow their descent through the atmosphere, also acted as a communications antenna, and a circular ring of shock absorbers below. Researchers packed a slew of new instruments into each of these landers, including a new camera system, a gamma-ray spectrometer to analyze surface rocks, and several meteorological instruments. The orbiters would allow the first sustained long-term observations of the atmosphere and its interactions with the solar wind. They would also serve as a radio link between the landers and Earth, eliminating the previous tricky necessity of sending signals directly from the Venusian surface to receivers here. The combined weight of the new Veneras, lander plus orbiter, was more than ten thousand pounds, the largest interplanetary craft ever constructed (see Figure 3.7).

3.7 The Venera 8 *spacecraft during prelaunch testing.*

Veneras 9 and *10* were launched four days apart in June 1975, from the Soviet launch complex at Tyuratam. Four months later the landers plunged into the atmosphere of Venus. Each one was equipped with instruments to measure the density of the clouds as they descended. It turned out that the clouds were not very dense and were relatively transparent, so that if you were in them they would seem more like fog or mist than like most clouds on Earth. But these diffuse clouds were also found to be incredibly vast, towering up to an altitude of forty-four miles from the cloud base at thirty-three miles. They were found to be concentrated in three discrete layers, with relatively clear air between, and each layer had a different mixture of droplet sizes, which hinted that they might be composed of multiple materials.

Venera 9 landed safely and photographed the surface of Venus. The first picture ever returned from the alien landscape of another planet* (shown in the insert) revealed a rough, sloping surface strewn with flat rocks extending towards a bright sky in the distance. The immediate impression, of an eroded field of volcanic rocks, has stood up under decades of scrutiny. *Venera 9* transmitted from the surface for fifty-three minutes before the signal was lost. Three days later *Venera 10* landed, at a site about twelve hundred miles away, also transmitting photographs of its surroundings, showing a similar landscape, with more rounded rocks and more dark, soil-like material between them. Did this indicate an older, more eroded site? What would erode and transport fine-grained material on a planet with no water and almost no surface winds? Each new observation produced many more unanswered questions.

"SEND IN A CLASS ONE PROBE!": THE VENUS PIONEERS

Poised for flight,
wings spread bright,
spring from night into the Sun.
—ROBERT HUNTER, "Help On the Way"

As the 70s wound on and Earth spun to a disco pulse, Venus—bright as a mirror ball but fuzzy as a shag rug—orbited sunward of us, presenting itself for Earthly visitation every nineteen months. At almost every opportunity the Soviets sent whatever they could scrape together. Their investigations proceeded tortoiselike while the American hare slept, dreaming dreams of Venus that turned to nightmares of congressional funding battles. American scientists went through several episodes of detailed planning and preliminary approval and funding only to be sent back to the drawing board by later eviscerating budget cuts.[†]

Studies for a next-generation Venus mission had begun in 1967, following the success of *Mariner 5*. Early ideas had included balloons that would float in the atmosphere and snazzy surface landers. Neither of these made the cut. Planners had optimistically assumed that each coming launch window would be used to send spacecraft to Venus.

---

*On October 22, 1975.

[†]The British science writer Eric Burgess, commenting on these difficulties and the consequent lost opportunities for American space exploration, asks: "What had happened to the American spirit of private enterprise and exploration that had so successfully developed a whole continent in less than two centuries and could have gone on to develop the Solar System?" He's got a point. If things had gone differently, by now we might have a McDonald's in Venus orbit, strip malls on the Moon, and MTV broadcasting "The Grind" from the canyons of Mars. Damn.

What survived was a more modest plan for two launches during the 1978 launch window. The flip side of all this was that by 1978, when the Americans were able to launch again, they were tanned and ready. Each stage of the frustrating, repetitive recasting (and descoping) of the plan had produced improvements in design. The resulting mission, *Pioneer Venus*, was less grandiose than some of the earlier plans, but it included many design innovations, and the scientists had time to cook up and refine a superb complement of new instruments that built on the results of previous missions. The whole *Pioneer* package was streamlined, efficient, extremely well integrated, and ripe with the promise of many answers and, of course, even more new questions.

The final configuration included an orbiter, launched in May, and a "multiprobe bus" launched that August (see Figure 3.8). The bus

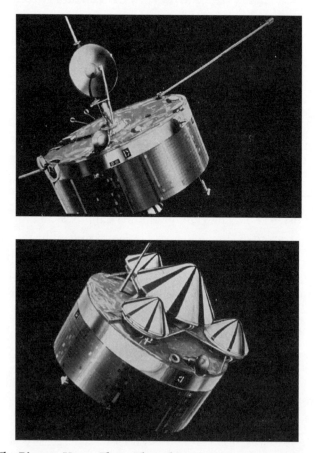

3.8 The Pioneer Venus Fleet: *The orbiter is on top. On the bottom is the bus with the four cone-shaped entry probes.*

was a cylindrical shell, covered with solar cells for power during the interplanetary journey. It would release four atmospheric probes with spherical vessels designed to carry instruments safely through the entire atmosphere in a protected low-pressure environment. There was one large probe, weighing seven hundred pounds, and three smaller, two-hundred-pound probes. Each was to sample a different location, latitude, and time of day as it descended. Earlier craft had revealed many of the basic characteristics of the atmosphere, including amounts of major gases, basic temperature, and pressure structure, and had placed rough bounds on the location and thickness of the cloud decks. Armed with this information, the designers of *Pioneer* knew what they were getting into, at least more than previous Venus explorers had. They had a basic knowledge of the conditions and physical demands that would greet the spacecraft at Venus, and they were able to design spacecraft to operate well there. The construction of these probes for the high-pressure lower atmosphere borrowed from experience designing bathyspheres to probe the depths of Earth's oceans.

A lot of hopes rode to Venus along with the scientific instruments on *Pioneer.* We expected this spacecraft to provide answers that touch on the grand questions that drive us to study the planets, questions about origin and evolution, the similarities and differences among worlds, and the uniqueness of the world we call home. In particular, the *Pioneer* scientists wanted to know why conditions on Venus are so different from those on Earth. To crack this puzzle, or at least delve deeper, they needed more precise measurements of the amounts of major gases, and an accurate census of trace gases and their variations with altitude. They needed to know at what levels solar energy was being absorbed and what was absorbing it. This could help provide the answer to the enigmatic superrotation. We also needed more precise measurements of how wind speed, temperature, and atmospheric turbulence varied with altitude. And were there horizontal temperature differences deep in the atmosphere that could help drive the intense high-altitude winds?

Other questions involved the great towering, diffuse cloud system. Does the layered structure discovered by the *Veneras* extend uniformly around the entire planet, or does this vary in space and time in response to atmospheric motions? How opaque are the clouds to different forms of radiation? Are they pure sulfuric acid or are there other materials mixed in? Are the different layers made out of different stuff?

Clues were badly needed to answer some important evolutionary questions as well. Had there ever been more water on Venus? Could it have had oceans? Has Venus always been hot as hell or has there been climate change from a milder (or even hotter) past?

And then, of course, there was the surface. If not a dripping jungle, what should we find there? All we had so far were a few intriguing snapshots and some preliminary chemical data provided by the Soviets. Was the surface broken up into continents and low basins like Earth? Were there mountain ranges and volcanoes? Does Venus have continental drift and seismic activity?*

Pioneer instruments could at least take a stab at each of these questions. The orbiter carried a radar altimeter that would create the first topographic map of the cloud-shrouded planet's surface. This worked by continuously sending radar pulses to the surface and precisely measuring the time delay of the return echoes. When the spacecraft was over a high-altitude spot, the echo took less time to bounce back. This all got fed into a computer that turned it into a topo map. The orbiter also carried spectrometers to study the clouds and upper atmosphere.

The one large entry probe contained several instruments designed to sample the atmosphere frequently during descent and make precise measurements of gas composition. Other experiments would simultaneously measure the radiation at several wavelengths and from several directions during descent. These latter experiments needed windows to look out of. But how to build a window that would let light through but hold off the crushing Venusian pressure and resist chemical attack by the acid clouds? No known glass would do, so the probe was sent to the surface of Venus bearing gifts of cut sapphires and diamonds.

The three small atmospheric probes also had devices to measure the changing light during their descents at different locations around the planet. The bus itself, destined to crash and burn in the upper atmosphere, carried a couple of spectrometers for additional sampling on its way in. All four entry probes also carried instruments to take profiles of cloud thickness as well as sensitive accelerometers to record spacecraft motions during descent and reveal subtle changes in atmo-

*Mango groves, bowling alleys, and Quickee Marts could probably already be ruled out on the basis of existing data.

spheric pressure with depth. This would help fill in the three-dimensional picture of the atmosphere and clouds. Scientists hoped to gather enough data to construct realistic models of the relationships among radiation absorption, chemistry, atmospheric motions, and clouds, providing the beginning of an understanding of how Venus works.

As December 1978 approached, Venus had passed near to Earth at inferior conjunction and was receding. It was a bright morning star in Earth's predawn sky. The two *Pioneer* spacecraft neared their rendezvous with Venus, the orbiter and the bus with four entry probes clinging tight to its underbelly like baby possums. And they were not alone. The inner solar system between Earth and Venus was rather crowded that December, as two giant Soviet spacecraft, *Veneras 11* and *12*, each equipped with a fly-by craft and a new, improved surface lander, were also making their way inward toward Venus. The Soviets had actually skipped the previous launch opportunity in order to study the large amounts of data acquired by *Veneras 9* and *10* and assimilate this into later designs. These next-generation *Veneras* included new, improved cameras for better surface panoramas and new instruments for measuring atmospheric gases.

And so it was that in December of 1978 the heavens of Venus were stormed by an armada of ten spacecraft from Earth.* On December 4 the *Pioneer* Venus orbiter successfully entered orbit around the planet, where it would continue to operate and take data for the next fourteen years, far outlasting anyone's expectations. Five days later the four *Pioneer* entry probes (plus the bus that had carried them) shrieked at twenty-six thousand miles per hour into the upper atmosphere where, having survived the rigors of interplanetary travel, each began its perilous journey down to the surface. Just twelve days after this, on December 21 (winter solstice in the North End of Earth), the *Venera 12* lander descended toward the surface of Venus as its companion fly-by craft relayed its signals homeward. And, four days later, on a freezing Christmas day in Moscow, the *Venera 11* lander, following close on the heels of its sister craft, dove at Venus and plunged toward the blistering surface.

All of these spacecraft worked, although all experienced some problems. The measurements made in that busy December nineteen

---

*Counting entry probes and orbiters as separate craft.

years ago still form much of the basis for our current knowledge of the atmosphere of Venus.

BRIEFING FROM A DESCENT INTO HELL

Let's follow the paths of the *Pioneer* entry probes during their demanding journeys from outer space to the surface of Venus. At 10:45 Pacific Standard Time, on the morning of December 9, the large probe screamed into the tenuous upper Venusian atmosphere, a glowing meteor sent from Earth. In just thirty-eight seconds air resistance slowed the craft from its entry speed of 26,000 miles per hour to 452 mph. During this phase the entry probe was engulfed in a ball of fire that shut off radio talk for about ten seconds, so measurements were stored in onboard memory to be sent home later. Then a parachute opened to slow the craft further, and its cone-shaped heat shield, needed to survive the entry, was jettisoned to allow all the instruments to "see" the atmosphere for the rest of the descent. After forty-three seconds in the atmosphere the spacecraft had descended to forty miles above the surface, and all instruments were radioing results home to Earth. During this brief plummet through the clouds, an unfortunate random accident occurred that didn't threaten the spacecraft in any way but screwed up some of its precious measurements: a tiny gas inlet, designed to sample the air repeatedly during descent, sucked in a cloud droplet, which became lodged there. This drop quickly evaporated, allowing more air to flow into the waiting instruments, but all subsequent measurements were now recording some unknown mixture of pure Venusian air contaminated with gases evaporating from the cloud droplet, making analysis much more difficult.

Eighteen minutes into Venus, the probe, having plunged through the sulfuric acid clouds, frantically gathering data along the way, had descended to twenty-eight miles above the surface. At this point the parachute was abandoned, so the rest of the journey was a free fall through the increasingly dense atmosphere.* The spacecraft continued

---

*The entry probe designers had to walk a thin line between falling too fast through the atmosphere, which would not allow enough time to take measurements during descent, and falling too slowly, which would cause the spacecraft to heat up too much and die before reaching the surface.

to plummet, sampling and examining the air every few seconds, slowed only by air resistance, until, thirty-six minutes later, it smashed into the surface, near the equator during daytime, traveling at about thirty miles per hour. The entry probe ceased functioning, or at least communicating with Earth, immediately on impact.

The three small probes each entered the atmosphere within ten minutes after the entry of the large probe. One minute after entry, each began to open its windows and doors to examine the clouds. The small probes were simpler in design and had no parachutes. They were given the names "Day," "North," and "Night" to reflect their entry points, which had been chosen to sample as diverse a range of latitudes and lighting conditions as possible. These three were all expected to die on impact like their larger companion, but one, "Day," survived and continued to function for sixty-eight minutes on the surface. The dying spacecraft dutifully transmitted its ever-increasing internal temperature until it reached 260 degrees Fahrenheit and finally succumbed to Venusian fever.

The feast of information returned by the entry probes raised many questions. One that has never been answered satisfactorily is "What the hell happened at 12.5 kilometers?" Each probe went haywire as it passed through a height of about 12 kilometers, or 7.5 miles, above the surface. The temperature and pressure sensors sent back crazy numbers, power surged throughout the probes, and some instruments stopped functioning entirely. Was this due to some still-unknown phenomenon that exists on Venus at this height, or was there some component, common to all four spacecraft, which failed quickly and unexpectedly at this altitude because of some environmental factor, temperature perhaps? To this day we do not have a widely accepted explanation for these nearly simultaneous freak-outs at the same altitude on four entry probes. In 1993, NASA held a workshop to review the data on, and possible explanations for, these "12.5 kilometer anomalies," but we still don't know what happened.

During its hour of glory, plunging from the frigid vacuum of space to smash down onto the sizzling surface, each *Pioneer* probe assembled a travelogue of data on the Venusian atmosphere that scientists still pore over today. Somewhere on the surface of Venus (we know the approximate locations) rest the remains of these crashed probes, which, like their Soviet counterparts, lie melting, frying, corroding, and eroding as they are gently buried by the ever so slowly shifting sands of Venus.

Even as the four *Pioneer* entry probes fell toward the surface, the bus, which had carried and protected them during their seven-month journey across interplanetary space and delivered each to its spectacular entry and demise at Venus, itself closed on the planet. The bus had hung back while the probes raced ahead of it, like eager children who just couldn't wait to get to Venus. Now, having lost its primary raison d'être as caretaker and transport for the entry probes, the bus made its own kamikaze dive. About a half hour after the last of the entry probes hit the surface, and as the "Day" probe still lay dying and frying below, the bus dove into the upper atmosphere. Lacking a heat shield to protect it from the intense heat of an atmospheric entry, it burned and disintegrated in about two minutes. As a Klingon Warrior would say, the bus "died with honor." Its instruments radioed home to Earth valuable measurements of the upper atmosphere and ionosphere, a region skipped by all the other probes, during its brief flaming streak through the southern daytime sky of Venus.

MEANWHILE, BACK IN ORBIT ...

With the end of the bus's brief flash of glory, just ninety-eight minutes after the excitement had begun with large probe entry, five of the six *Pioneer* spacecraft had completed their missions. As the charred fragments of the bus wafted slowly down through the atmosphere, the mission focus shifted to the one remaining craft, the orbiter. The orbiter was much longer lived than the entry probes, and its pace of discovery was less frantic, more measured. Yet it had begun, a few days before probe entry, with considerable excitement and anxiety.

To end its interplanetary flight and enter an elliptical orbit around Venus, the orbiter's large rocket had to fire and then shut down at precisely the right times. Unfortunately, the flight geometry was such that this maneuver had to occur when the spacecraft was on the side of Venus opposite Earth. Thus the commands for these operations could not be sent directly, but had to be stored in the spacecraft's onboard computer memory. During the flight from Earth, this computer had proven to be somewhat flaky. It required corrections from

Earth every couple of weeks when a cosmic ray zap would randomly switch zeros and ones in its binary memory. We'd lose the orbiter if something like this happened during the seven-minute gap between the time when the spacecraft passed out of sight and the time the rocket motor was to be fired. There would be no second chance. To compound the anxiety, no one had ever fired a large rocket motor of that type after such a long space journey. The orbiter would have to make these unprecedented maneuvers on its own. The critical moment arrived, and for thirty seconds the mighty little engine roared, slowing the spacecraft by about twenty-three hundred miles per hour. The scientists and spacecraft operators at NASA's Ames Research Center along the southwestern shore of San Francisco Bay cried out with joy and relief minutes later when they received the first signals. *Pioneer* had successfully achieved orbit. Venus had a new Moon.

Smaller thrusters were fired to fine-tune the orbit. Once it was adjusted to the right shape, the instruments were turned on and data started to stream homeward. The methodical topographic mapping by radar began.

Using the ultraviolet markings as atmospheric tracers, the orbiter measured wind speeds at the cloud tops to complement wind-speed measurements made by the entry probes at different altitudes. Our picture of global circulation patterns began to clarify. With each orbit, the radar altimeter scanned a narrow strip of the planet's surface. Gradually a tantalizing map of the Venusian surface, the first ever, began to emerge. The partially completed topographic map revealed large, continental-scale mountainous areas and great flat rolling lowland plains. This immediately seemed familiar to observers from Earth; a planet divided up into continents and lowland (ocean) basins. If there were large impact basins* (which would indicate an ancient surface like the Moon, Mercury, or Mars), they didn't jump out at us.

Before the mapping was complete, an alarming glitch occurred: the altimeter simply stopped working. No one could figure out what the problem was and, with great regret, the *Pioneer* scientists shut off the instrument. But when they turned it on again a month later, it worked

---

*An impact basin is just a huge impact crater. When a crater is large enough to dominate the geography of an entire region of a planet's surface (hundreds or thousands of miles across), we call it a basin.

just fine. They determined that if they kept the radar on for short periods, ten hours or less, and then gave it a rest, it would not overheat and shut down. From that point on the instrument was operated for only a short time on each orbit. Mapping continued in this mode without further problems. A large gap appeared in the first maps, where we had lost that month of data, but we were able to cover this ground later in the mission when the spacecraft again orbited over these same areas. This map is shown in the color insert. This is just like the topo maps you take hiking, except that colors instead of contour lines are used to represent different heights and the smallest "bumps" you can see are about fifty miles across. This is the size of the *Pioneer* radar "footprint," the smallest surface feature it can distinguish. It is also referred to as the "resolution" of the map, and this low resolution, more than anything else, limited our ability to interpret the map. As you can see, Venus has a couple of very large elevated "continents," a handful of smaller high-altitude areas, and vast regions of smooth lowlands.

Spirited debates immediately began on how (and whether) these features fit into familiar terrestrial analogies. Were the two large elevated areas, Africa-sized "Aphrodite Terra" along the equator and Australia-sized "Ishtar Terra" near the North Pole, really continents, in the terrestrial sense of being composed of less dense rocks like granite that floated on denser, interior rocks? On Earth, large-scale differentiation into such areas of dramatically different rock types is the result of plate tectonics and convective motions and melting in the planet's interior. So finding true continents on Venus would have exciting implications for the internal evolution of our sister world. The creation of the *Pioneer* topo map raised the prospect of doing comparative planetology with unprecedented relevance for understanding the interior and large-scale evolution of Earth.

The *Pioneer* orbiter circled Venus for nearly fourteen years, gathering data all the while, until October 1992, when, out of fuel and losing altitude, it met a fiery demise in the upper atmosphere. Operators on Earth kept contact up to the last possible moment, and the orbiter sent back crucial new information on the upper atmosphere with its last electronic breath.

The major findings from the *Pioneer* and *Venera* probes were collected in a book called the *Venus International Reference Atmo-*

sphere.* Although not exactly a best-seller, it is a cherished reference among students of Venus's atmosphere, and many a copy has become dog-eared and worn. The tables and summaries of atmospheric data found therein are still the standard on Earth for Venus models, and the wide use of this standard allows us to make sure that we are comparing apples with apples, when making models and sharing new results.

So, what did we find there? What did we learn? Too much to describe here, of course. Papers describing and analyzing the data from *Pioneer* and its Soviet contemporaries fill volumes. But I will try to briefly abstract some highlights from this important time of discovery.

STRANGE BREW: WHAT'S IN THE AIR DOWN THERE?

Science ain't an exact science.
—FROM THE FILM *Twelve Monkeys*

One thing we hoped to learn with *Pioneer* was the identity of the unknown ultraviolet absorber, the strange stuff that makes the dark ultraviolet streaks and swirls in the clouds, whipping around the planet with the superrotating atmosphere. Planetary atmospheres are driven by the Sun. Sunlight powers their motions and much of their chemistry. We wanted very much to know what the dark stuff is because it is one of the main absorbers of solar energy on Venus and thus seemed key to unlocking many mysteries. We didn't actually succeed in naming the mystery absorber with the data from *Pioneer,* but we were able to learn much more about it, to pin down its location, some of its physical properties, and patterns of movements.

Just as you can watch the Sun disappear as you descend through clouds in an airplane, *Pioneer's* radiometers measured the spectrum and intensity of the fading sunlight as they fell into the clouds of Venus. They found that a huge amount of solar energy was absorbed at near-ultraviolet wavelengths,[†] in a pattern matching that of the

---

*Published by the Committee on Space Research of the International Council of Scientific Unions.

[†]*Near-ultraviolet* means "just outside the visible range."

unknown absorber at altitudes in and above the upper layer of clouds. The distribution and motions of this stuff are very closely tied to the presence of SO_2 (sulfur dioxide). It seems clearly related to the chemistry of sulfur molecules in the upper atmosphere, driven by UV light. Planetary scientists have proposed and reluctantly discarded many candidates for the identity of this mysterious and powerful UV absorber, as experiments in laboratories on Earth have shown that the candidates don't measure up to the observed properties. The mystery remains.

Speaking of sulfur (weren't we?), *Pioneer* found large amounts of sulfur-containing gases, mostly SO_2, throughout the atmosphere of Venus. These gases are much more rare on Earth and are associated mainly with volcanic emissions or smokestack pollution. On Earth, SO_2, broken into radicals* by ultraviolet light in the upper atmosphere, reacts with water to create acid rain. This is becoming a serious problem in many areas, decimating forests, fish, and frogs, and chemically eroding Venus glyphs from ancient Mayan buildings. On Venus, these same chemical reactions create the global cloud decks of concentrated sulfuric acid. The acid rains of Venus never make it even close to the surface, forming instead a zone of perpetual virga† beneath the clouds. Since acid rain is the only kind of rain there is on Venus, I suppose it's not a problem there. The extreme case of relentless acid rain on Venus is one of many ways in which our neighbor planet turns out to be a natural laboratory for studying some of our most pressing environmental problems on Earth.

Sulfur on Venus takes many different forms. The data from *Pioneer* hinted at the existence of a complex sulfur cycle on Venus, involving the atmosphere, the clouds, surface minerals, and perhaps even the interior of the planet. This is very exciting for comparative planetology. One of the things we have been realizing, more and more, about Earth, is how its workings are characterized by complex

*A radical, in chemical terms, is a molecule fragment that is highly reactive. Radicals are formed when some energy source, often ultraviolet radiation, rips apart the chemical bonds of stable molecules. Radicals are unstable, incomplete. Radicals are unhappy with their state of being and will do what they can to change things, sometimes affecting innocent molecules who happen to be in their way.

†This is the word for rain that evaporates before it reaches the ground, common on desert areas of Earth.

cycles of carbon, oxygen, and other elements through the atmosphere, surface, oceans, and interior. We have not been able to watch any active chemical cycles of similar complexity occurring on other planets. But maybe sulfur on Venus qualifies as a genuine global chemical cycle, rivaling Earth's in complexity. If we can understand the sulfur cycle on Venus, maybe we can perceive the vital chemical cycles of Earth in a less provincial light.

On Venus, sulfur is a chemical element of remarkable promiscuity. Every planet seems to have one. Like oxygen on Earth, and hydrogen on Jupiter, sulfur on Venus couples with nearly anything for the right exchange of energy or just to achieve electronic satisfaction. In brief, from top to bottom (above the clouds down to the surface), the parts of this cycle are as follows (see Figure 3.9).

Above the clouds there is a *photochemical zone* where solar ultraviolet light is the provocateur, providing the energy to drive chemical

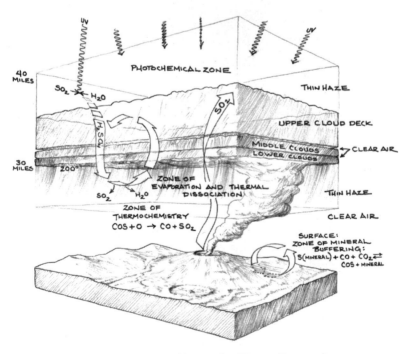

3.9 The Venusian Sulfur Cycle. *(Carter Emmart)*

reactions. UV zaps apart molecules, and the resulting fragments (radicals) scramble like mad to reassemble themselves into new stable forms. This drives SO_2, water, and other chemicals, to react, forming sulfuric acid (H_2SO_4). This process is exactly analogous to the formation of a "photochemical smog," with which we are all too familiar. Ultraviolet light acts on gases emitted from our smokestacks and car exhaust pipes to produce the dreaded "brown cloud," like the one I can see hovering menacingly over Denver right now.

This is also the region where the unknown ultraviolet absorber is doing whatever it does, apparently also involving sulfur.[*] The top of the clouds is defined by a level where the sulfuric acid condenses into tiny droplets, which grow and fall out of the photochemical zone. This zone ends at a depth where most of the ultraviolet light from the Sun has already been absorbed. As the droplets continue to fall through the clouds, they grow in size, both by colliding and merging with one another and by condensing sulfuric acid and water vapor out of the air around them. They fall into ever-hotter air as they approach the scorching surface.

After falling more than twelve miles, the swollen droplets reach an altitude of about thirty miles, where the temperature is 200 degrees Fahrenheit. Here we find a *zone of evaporation and thermal dissociation*. At this temperature, the liquid drops of concentrated sulfuric acid can no longer contain themselves, and they start to boil. The molecules have enough thermal energy (heat) to break free of the liquid state and rejoin the great ocean of air around them. The drops evaporate, and molecules of sulfuric acid dissociate (rip apart) in the heat, breaking back down into SO_2 and H_2O gases.[†] (Since this is a cycle, there is an equal number of upward moving sulfur atoms here doing the opposite: forming H_2SO_4 molecules and condensing into liquid droplets.)

Descending farther in altitude from this point, it just gets hotter. Here is a zone dominated by *thermochemistry*[‡]: it is so hot that the gases don't need radiation or lightning or any other inducements to

---

[*]You could look at this stuff as Venus's ultraviolet sunburn!

[†]In general, the hotter it is, the less stable large molecules are.

[‡]Thermochemistry is, essentially, just *cooking*.

react. They do it all by themselves because at these high temperatures the molecular motions become so ferocious that the energy of random collisions is sufficient to cause reactions. Atmospheric pressure down here is significantly higher than at sea level on Earth, so collisions are more frequent, further facilitating vigorous chemistry. In this zone, sulfur atoms pass rapidly between several different kinds of molecules, all of which we would find stinky and toxic.

The next zone is down at the surface. Here gases from the atmosphere react with minerals in the rocks. This is the zone of *mineral buffering*, so called because the chemical composition of the surface rocks can control or "buffer" the gas composition of the atmosphere. It's so hot that atoms exchange easily between gas and rock. The chemicals in the rock have a strong effect on the mixture of gases in the air. Sulfur in the surface rocks reacts with CO_2 and CO in the air to make an odd little molecule called carbonyl sulfide, or COS. It is also likely that down at the surface sulfur gases, probably mostly SO_2, are being breathed into the atmosphere by active volcanoes. This is one of the ways that the chemistry of the planet's interior regions may participate in this huge cycle of sulfur.

We have good reasons to suspect that active volcanoes are currently supplying large amounts of sulfur gases into the atmosphere. In the lower atmosphere *Pioneer* found an amount of SO_2 too high for reactions with surface minerals to be controlling its abundance. This is evidence, albeit somewhat indirect, for an ongoing volcanic source of this gas. Thus volcanoes may be the ultimate source of the sulfur that makes the clouds. The clouds, which so dominate the appearance of the planet and the atmosphere's response to sunlight, might disappear if the volcanoes stopped. It seems that activity at the surface has been the culprit all along in maintaining those planet-obscuring clouds. The surface hides itself from extra-Venusian scrutiny by putting up a global smoke screen, with sulfur as a major conspirator.

One controversial observation made from the *Pioneer* orbiter may be direct evidence for such an eruption of volcanic gases. The ultraviolet spectrometer on the orbiter detected a steadily decreasing amount of SO_2 above the clouds, over a time span of several years. Why was the SO_2 disappearing right before our eyes? Maybe a giant volcano erupted shortly before *Pioneer* arrived at Venus, blasting a cloud of sulfur into the upper atmosphere. This happens frequently on Earth, when large volcanic eruptions pump SO_2 into the stratosphere where ultraviolet light turns it into sulfuric acid droplets, in chemical reactions that are essentially identical to those occurring in the cloud tops of Venus. These volcanogenic acid hazes temporarily cool Earth slightly by blocking sunlight, and make colorful sunsets by scattering sunlight, but they last only a year or so because the sulfuric acid is rapidly "rained out" by Earth's ubiquitous water cycle. Venus has no water cycle to speak of, so any such volcanically generated sulfur haze would stick around longer. Some researchers saw the declining level of SO_2 above the clouds as the lingering "smoking gun" of a giant eruption, proving that Venus had active volcanoes of a very large and explosive variety. Others argued that the changing SO_2 levels might just be part of a long-term cycle of atmospheric circulation that mixed SO_2-rich air from below the cloud tops into the high atmosphere. The jury is still out on this one.

And so the atoms of sulfur go round and round, participating in every part of the life of the planet, jumping from one chemical compound to the next, each over millions of years doing time in minerals, magmas, gases, and liquid cloud drops, from down in the mantle up to the hazes above the clouds.

WHY CHEMISTRY?

Here perhaps I should say something about why chemistry happens at all. This should be of interest to you, whose body and mind can be described largely, if not completely, as a complex system of chemical reactions.

Why do molecules react, interact with one another, and trade atoms to become new molecules? You can blame this on the atoms themselves, their overriding need to arrange their electrons in certain configurations, to reach a lower energy state, to achieve stability.

Remember, the atoms of each element have a characteristic number of electrons, equal to the number of protons bound within the nucleus, and it is this electron cloud that gives each element its identity. Tiny hydrogen has only one electron (and one proton).* Carbon has six, and the largest atoms have more than a hundred. These electrons are arranged in various shells around the nucleus, and what gives each element its *personality*, defining its interactions with other elements, is the number of electrons it has in its outer shell. This is what each element has in common with others in its column in the "periodic table of the elements" that you've seen on the wall of science classrooms. Right below H in the periodic table you see Na (sodium). These elements enter into many of the same configurations because they both have one electron in their outer cell. So, for example, you get HCl (hydrochloric acid) and NaCl (table salt) when these elements pair up with chlorine.

The way an atom achieves better living through chemistry is to fill up its outermost electron shell, by borrowing, lending, or sharing electrons with other atoms. A full outer shell contains eight electrons. This is what every atom wants, and they need each other to attain it. This is what drives atoms to conspire together to make molecules. Chlorine has seven electrons in its outer shell, and thus is only one electron short of full. This is why Cl loves to mate with H and Na, each of which has one electron to spare, making the familiar compounds mentioned above. Thus is made the salt of the Earth. The only atoms above all this socializing are those that already have full outer shells—the noble gases, helium, neon, argon, krypton, and xenon, which occupy the far right column in the periodic table. They don't need anyone else; born in a highly stable state of electronic nirvana, they remain celibate, unreactive. All other elements behave as though they are striving, through their electronic interactions with one another, to mimic the full outer electron shells of the noble gases.

Everything in and around you—the cellulose in the pages of this book, the proteins in every cell of your hand that holds the book, the oxygen and nitrogen molecules in the air you are breathing, and the aspirin you will need after reading this section—is held together by chemical bonds formed according to these rules.[†]

---

*What hydrogen lacks in size it makes up in numbers. Most of the universe is hydrogen.

[†]The exception is the argon in the air which, due to its "nobility", doesn't form bonds with anything, not even with itself. We've all met people like that.

One reason for sulfur's prominence on Venus is the very high surface temperature. This favors chemical reactions that release sulfur from surface minerals into atmospheric gases. But another important reason for the reign, and the rain, of sulfur on Venus is the remarkable lack of water. On Earth, water is a scavenger, keeping the air clear of sulfur and many other gases that dissolve in raindrops. On Venus, with such a dry atmosphere, water cannot play this role and sulfur has the run of the place.

You may notice that this is a recurring theme. Sulfur rules on Venus, but water rules on Earth. The more we understand, the more it seems that most differences between these twin planets may boil down (so to speak) to the huge difference in the amount of water. An important goal of the *Pioneer* mission was to determine the amount of water in the lower atmosphere, below the clouds, a goal that had long eluded us. We had known for a long time that the above-cloud atmosphere was remarkably dry, and that the clouds were made of acid instead of water. Both facts strongly suggested that the planet as a whole was quite desiccated. Certainly, a surface "hotter than Georgia asphalt" would have no liquid water, only a trace amount of water vapor in the air. But how much of a trace? We had already performed many experiments designed to answer this question, but the answers seemed flaky because they were all over the map. I am not talking about minor discrepancies: some studies yielded water amounts *hundreds* of times greater than others.

The *Pioneer Venus* large probe carried an instrument called a mass spectrometer that measured the molecular weights of gases in the atmosphere and determined their amounts. Scientists hoped this instrument would help clear up the water puzzle. But, in a stroke of bad luck, that pesky cloud droplet got sucked into its intake during the descent. The mass spec continued to take data, recording gas composition, but of what? The investigators on Earth tried heroically to recover useful data from the confused tangle of evaporating cloud vapors mixed with pristine atmosphere. Unfortunately, these data will always be a little bit suspect. The numbers they came up with were as follows: one hundred parts per million water below the clouds decreasing rapidly with altitude to 20 parts per million near the surface. This didn't clarify the situation much: the results were consistent with some other experiments and inconsistent with others.

To make matters worse, or at least stranger, it made no sense that the water abundance decreased with altitude. The lowest part of a planet's atmosphere is almost always well stirred. Atmospheric motions (weather) mix the gases together and you generally don't find *compositional gradients.* A gradient is a systematic change in concentration across space.*

Pioneer's report of a water gradient, getting drier toward the surface, seemed wrong. There were three ways out of this paradox:

1. Many people simply did not believe the data, preferring to believe that the clogged inlet caused an erroneous result.

2. The steep gradient might be explained by a rapid flow of water through the atmosphere. Why? Because a gradient often implies a flow.

Let me explain: If the measured concentration of some ingredient changes across a certain distance, it often means that materials are flowing through the system. In the absence of a flow, random motions will wipe out any gradients, leading to a well-mixed atmosphere. Think of a river delta or coastal estuary where fresh water from a river is running into an ocean of salt water. The water changes gradually from fresh to salty as you head out to sea from the mouth of the river. This *gradient* of salinity is caused by the continual flow of fresh water. If the flow stopped, the gradient would quickly disappear as the water was stirred and mixed by the ocean currents. A flow maintains a gradient against the natural tendency for random mixing. A gradient implies a flow.

The gradient observed on Venus could be explained if water were somehow being poured in beneath the clouds and removed at the surface. It sounds good in theory, but the numbers don't work out. A simple calculation to see what rate of water flow you need to get the observed steep gradient gives you absurdly high numbers. The surface would have to be sucking up water at too high a rate to be maintained for more than a geological instant. Why should we happen to be observing at that instant, and what would be the source of water gushing from

---

*For example, as you go west across North America, there is a *gradient* of people who begin every sentence with "Dude": the number increases as you approach California.

the clouds to the surface? So this idea was often mentioned but seldom advocated.

3. Maybe water was being chemically converted to some other hydrogen compound that was more stable in the hot lower atmosphere. Stability is everything in chemistry. There is no loyalty, and hydrogen would jump ship in an instant if it could find a more stable configuration. A few such chemical schemes were proposed, but none was supported by any other evidence, and so they never gained wide acceptance.

To add to the confusion, an elegant experiment included on several of the Soviet *Venera* landers seemed to confirm the strange water gradient. Using instruments called *spectrophotometers*, the Soviets had measured the spectrum of fading sunlight as the landers fell deeper and deeper into the atmosphere, watching for the telltale absorption signature of water and other compounds. These results resembled those given by the *Pioneer* mass spectrometer, although some of the details looked different. However, some scientists carefully reanalyzed these data and concluded that the absorption seen higher in the atmosphere, usually blamed on water, was caused by some other, as yet unknown, gas and that twenty parts per million might be a more accurate water amount for the entire lower atmosphere. This debate went on for over a decade (see Figure 3.10).

WHO CARES?: WATER ABUNDANCE

Why did we put all this intense effort into pinning down the exact amount of water in the atmosphere? After all, we already knew that it was pretty darn dry there, so wasn't that good enough?

Actually, knowing the water abundance as accurately as possible turns out to be of prime importance for many reasons. First of all there are some important questions of comparative planetology, for which water is key. Water defines the character of our own world, covering three-quarters of the surface with oceans, playing an essential role in Earth's climate, dominating erosion of mountains, and even affecting the properties and motions of the interior. And then of course there's water's essential role in the origin, evolution, and sustenance of life on Earth.

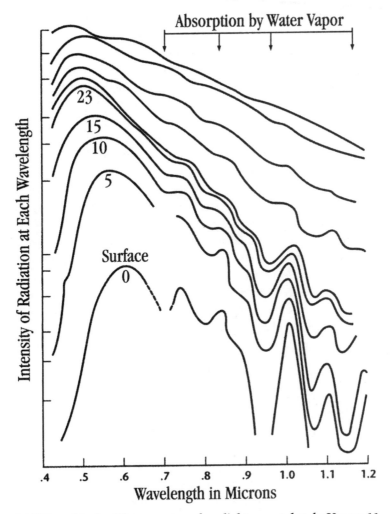

3.10 Water Marks: *These spectra of sunlight were taken by* Venera 11 *as it descended toward the surface. The number by each curve is the altitude (in miles) at which that spectrum was taken. The arrows show the wavelengths where water vapor absorbs light. Note how the light disappears at these wavelengths as the craft goes deeper into the atmosphere.*

But that's Earth. Weren't we discussing water on Venus? The two planets are similar in so many respects that the differences really jump out at us. Perhaps the key difference between these neighboring worlds, and one that may help to explain most of the others, is the striking lack of water on Venus.*

We want to understand where this difference came from and what the consequences are. One of our key avenues toward such understanding comes from making models. Models, in this sense of the word, are mathematical expressions of some simplified ideas about how a planet works. These days they are almost always created with computers. If a model is any good, it encapsulates and illuminates some important features and behaviors of the planet.

Two kinds of models for which the water abundance is an important variable are climate models and evolutionary models. On Venus, water is a trace greenhouse gas, just like carbon dioxide is here on Earth. So small changes in the water abundance can have important effects on the climate. To make good climate models for Venus, we need to know how much water there is. This not only helps us understand climate on Venus but also helps us test our ideas of planetary climate, which we use for predicting global change on Earth. This is one of the many places where our study of other planets crosses the line from the mere satisfaction of intellectual curiosity to the gaining of crucial survival skills.

Evolutionary models help us extrapolate from currently observed conditions on a planet to conditions in the distant past, stretching all the way back to the formation of all the planets from a common disk of debris surrounding the young Sun. In these models we develop equations to represent processes that may have added water to the atmosphere (volcanoes, icy comet impacts), those that have removed it (escape into space, chemical reactions with surface rocks), and changes in the rates of these processes over time.

Evolutionary models are our best bet for answering the vexing question "How did Earth and Venus get to be so different?"

These models are also only as good as the data that we put into them, and here again the current amount of water turns out to be critical for distinguishing between different possible planetary histories. For each of these purposes an order of magnitude

*Or overabundance on Earth. It's hard not to be geocentric in this comparison.

uncertainty is completely unacceptable,* because the models give very different results for an assumed water abundance of twenty or two hundred parts per million.

The *Pioneer* mass spectrometer data were far from useless, even though their interpretation was made difficult by the cloud droplet incident. Planetary missions are rare opportunities, and we are so starved for data that every bit is precious. (It would be worth much more than its weight in gold even if it did weigh something, which data doesn't.) The effort to make sense of this priceless, albeit somewhat confusing, stream of numbers continues to the present. Some reanalyses were published well over a decade after the probe's fifty minutes of fame.

An accurate determination of the water amount was one of the important goals of the *Pioneer Venus* mission. That more than an order of magnitude uncertainty remained after all those missions was a bit of an embarrassment, considering that we had sent not just one but several probes with sophisticated analytical instruments right into the planet. We had gone there, more than once, and demanded an answer, and all we got was a colossal riddle.

But this doesn't mean that *Pioneer* was a failure. In fact, as a whole, this mission was one of the most stunning successes of planetary exploration. To measure the amount of water in a planet's atmosphere is not easy. We are still not sure exactly how much there is in Earth's stratosphere! The persistence of the water mystery is a testament to the strangeness of the Venusian environment and the enormous difficulty of sending experiments to be performed there and interpreted here.

WAS VENUS WET?

Every cloud droplet has a silver lining, even giant ones made of battery acid that foul your once-in-a-lifetime opportunity to probe an alien

---

*A difference by one order of magnitude is a change by roughly a factor of ten: ten times larger or ten times smaller. An order of magnitude uncertainty means that the highest estimate of a quantity, in this case the water abundance, is roughly ten times higher than the lowest estimates.

atmosphere. The clogging of the mass spec inlet certainly was a set-back. But, like many disasters, it was also an opportunity when seen in the right light. Eventually this incident led to one of the most remarkable discoveries of the *Pioneer* mission. The guilty cloud drop, evaporating in the growing heat surrounding the descending probe, flooded the mass spec with the gas products of its own demise. This meant that a lot of excess hydrogen entered the instrument (all those H's in the evaporating H_2SO_4). Mixed in with the signature of this copious hydrogen was another, surprising signal indicating huge amounts of *deuterium.* This lucky find was recognized as a potentially vital clue for unraveling the divergent histories of Venus and Earth.

Deuterium is heavy hydrogen. Ordinary hydrogen is the simplest atom conceivable; one proton and one electron. However, the Big Bang blessed some hydrogen atoms with a neutron as well, and these atoms are what we call deuterium.[*] Since hydrogen and deuterium have identical electron structures, they behave almost identically in chemical reactions, where electrons rule. But deuterium is twice as massive, so any process that separates atoms by mass will discriminate between them. We call this mass fractionation, and because it treats hydrogen and deuterium differently, it can change the deuterium-to-hydrogen (D/H) ratio.[†]

What *Pioneer* found, courtesy of the randomly clogging drop, was that hydrogen on Venus is extremely fractionated; the measurements indicated a phenomenally high D/H ratio. Earth's oceans have an average of one deuterium atom for every six thousand hydrogen atoms. On Venus, we measured a ratio roughly 120 times greater. This means more than 2 percent of the hydrogen on Venus is of this odd, heavier variety.[‡]

How can we explain this massive buildup of heavy hydrogen on Venus? Clearly, some highly effective process that discriminates by mass has been at work for a long time on the planet, getting rid of

---

[*]From Genesis to Deuteronomy, my wisecracking Ph.D. advisor used to say!

[†]That is, the number of deuterium atoms divided by the number of hydrogen atoms.

[‡]There is an even rarer, heavier form of hydrogen with yet another neutron, called tritium, which is radioactive. If you want to build a hydrogen bomb, you will have to get hold of some of this.

hydrogen and concentrating deuterium. It seemed clear that the fractionating force must be simple gravity.

All planets with atmospheres are losing gases to space all the time. This escape flow is composed of the lightest gases and so is usually mostly hydrogen. It is one of the important ways in which the composition of planetary atmospheres changes, evolves, over time.

The ease of atmospheric escape is strongly dependent on the mass of the gravitational prisoners. The lightest atoms get accelerated to the highest speeds by the occasional collision in the low-density gas mixture of an upper atmosphere. The same amount of random collisional energy will launch a hydrogen atom to a velocity four times higher than it will an oxygen atom. In the mosh pit of the upper atmosphere, the little H dancers get thrown around a lot more than the comparatively huge O's. This is why hydrogen dominates the escape flow. Those fast-moving atoms that happen to be heading up, away from the planet, will escape into space, leaving Venus for good, to join the wispy interplanetary hydrogen winds.

A steady stream of hydrogen is escaping from the top of Venus's atmosphere at present. Observations from the *Pioneer* orbiter were key in helping to pin down the rate of this escape and the physical processes responsible for it.

Since deuterium is twice as heavy as hydrogen, it has a harder time escaping. As H escapes and D is left behind, the percentage of D in the remaining atoms goes up. The more hydrogen escapes, the higher the D/H ratio. The implication was clear: a very large amount of hydrogen had escaped into space from the top of Venus's atmosphere throughout the planet's history.

Given the dramatic contrast in water abundance between the planetary "twins" Venus and Earth, the conclusion was irresistible: The discovery of the huge D/H ratio meant that Venus was once wet! As the hydrogen from Venus's vanishing water supply has escaped over the eons, it has left behind a residue of deuterium, resulting in an ever-increasing D/H ratio. This discovery was hailed as the long-sought direct evidence for the "missing" oceans of Venus.

The very high D/H ratio allowed us to extrapolate back from the current dry state of Venus to a glorious Earthlike past with oceans and a more moderate climate. The first researchers to publish reports of this discovery, and its possible implications, may have been a bit too eager to find proof of just such a history, perhaps reflecting a desire to return somehow to the Venus we (thought we) knew and loved before *Mariner 2*, the Carboniferous Venus with oceans and primitive creatures, a friendly home in the sky.

But, there were two problems with the initial "Venus was wet" interpretation of the D/H ratio. First, it involves an assumption about the original D/H on Venus. A wide range of D/H values are observed throughout the solar system, and the origin of Earth's value is not well understood. So, the assumption that the initial D/H on Venus was identical to the current terrestrial value, while representing one reasonable possibility, should not be used to draw definitive conclusions about the history of water on that planet. To assume that this terrestrial "standard" holds currency elsewhere, without a fuller understanding of the origin of this value, is somewhat geocentric.

The second problem with the interpretation of the D/H ratio is that it assumes that the water abundance has simply been declining over the lifetime of Venus. It does not allow for the possibility of ongoing sources adding hydrogen back into the atmosphere. The assumption was that hydrogen is only escaping, not being supplied in any way, and that Venus has been on a steady one-way path from a more Earth-like waterlogged past to its current desiccated state. In this scenario, the H escape occurring now is the last trickle of a continuously escaping water reservoir that may have begun with a mighty ocean. However, other interpretations are possible. If hydrogen, in water or any other form, is being added at a rate comparable to the escape, then the relationship between current deuterium and past water is not so simple. The high D/H ratio *is* strong evidence that a lot of hydrogen has escaped over time, and it probably does mean that at times Venus had substantially more water than it does now. But saying that this observation is proof that Venus once had oceans just like Earth is a bit of a stretch.

One reason for thinking that Venus may not be losing water at present, but breaking even in the long run, is the short "lifetime" of water on Venus: if you let the amount of hydrogen contained in all the water that is now on Venus escape at the currently estimated rate, you would run out of water completely in a fairly short time, 100 to 200 million years. This is the lifetime, and it's considerably shorter than the age of the planet. It seems unlikely that we came along at just the right

A Mayan Observatory: *The Caracol in Chichen Itzá was custom built for Venus viewing around A.D. 1000. (Anthony Aveni)*

Thanksgiving Gathering over Denver Post Office: *Venus (at left) holds court on November 23, 1995. The guests are (from left to right) Mars, Jupiter, and a day-old crescent moon. (Tory Read)*

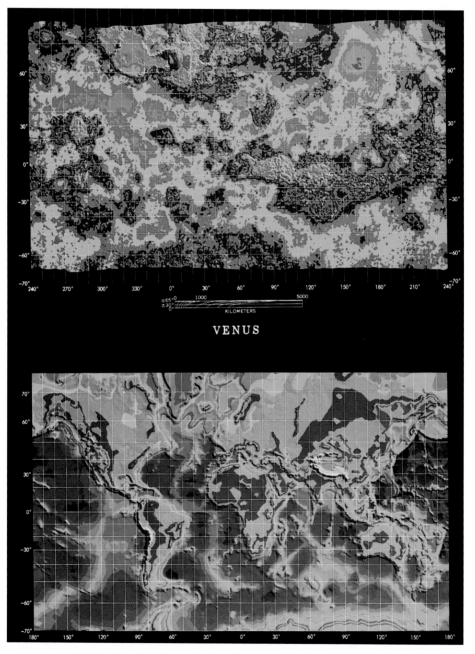

VENUS

Two of a Kind?: *The global topography of Venus and Earth. The map of Venus was made with the radar altimeter on the* Pioneer Venus Orbiter. *(USGS)*

Venera 13: *A mockup of the* Venera 13 *spacecraft on display at the Cosmos Pavillion in Moscow. The lander (foreground) rode to Venus inside the sphere atop the* Venera *spacecraft (background).*

In the Red: *This color image of the surface of Venus was made by the Soviet* Venera 13 *lander in March 1982. Filtering of light by the clouds and thick atmosphere results in the ubiquitous red color. The flat rocks are probably eroded basalt flows. Bright sky above the horizon can be seen in the upper right corner.*

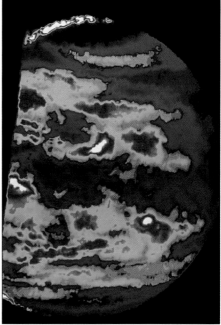

Two New Faces of Venus: *The top image was taken by* Mariner 10 *in February 1974. Ultraviolet images revealed detailed dark markings in the cloudtops which circle the planet every four days, constantly shifting forms. The bottom image was made by the* Galileo *spacecraft in February 1990. This view of the night side through a near-infrared "window" (at a wavelength of 2.3 microns) highlights structures in the lower clouds, six to ten miles below the visible cloudtops. The white and red spots are places where the clouds are thinnest. (NASA)*

Bon Voyage: Magellan *is launched from the space shuttle* Atlantis *in May 1989. The main antenna is at top. The solar panels are still folded from the shuttle trip to low Earth orbit.* (NASA)

Global View: *These four hemispheric views were made by combining Magellan's images and topographic data. A global image mosaic was projected on a sphere, and color was added to represent topography. Features larger than two miles across can be seen. (USGS)*

Centered on 0 degrees longitude. Ishtar Terra is seen far to the north. The large white (elevated) area is the high peaks of Maxwell Montes. Alpha Regio is seen in the south, a large plateau of rugged tessera terrain.

Centered on 90 degrees east longitude. Western Aphrodite Terra is seen along the equator. The circular feature to the southeast is the giant Artemis Corona.

Centered on 180 degrees east longitude. The center of this hemisphere is dominated by the steep valleys of central Aphrodite Terra. Further east is the large volcanic complex Thetis Regio, dominated by two huge shield volcanoes, Maat Mons and Ozza Mons. Giant fractures can be seen radiating outward across the plains from Thetis.

Centered on 270 degrees east longitude. The elevated area in the northern mid-latitudes is Beta Regio.

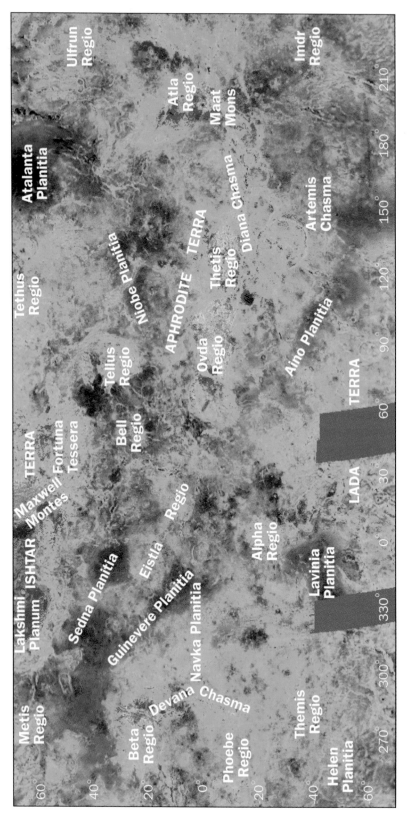

Magellan's Topo Map: *This color-coded topography map was made with the radar altimeter on board Magellan. The colors from highest to lowest altitudes are red, yellow, green, light blue, and dark blue. The resolution (or footprint size) is six miles near the equator and as much as fifteen miles at higher latitudes. Gray areas show the gaps in Magellan's coverage. (NASA)*

instant of cosmic time to observe the last few molecules of a former ocean flee into space. The short lifetime suggests (but does not prove) that a supply of new water is feeding the escape flux. Rather than seeing it as in constant decline, a better way to model the current water budget of Venus might be a "steady-state" condition, which means that some source of water is balancing the "sink" of escape.

Imagine this: You show up at a house where no one seems to have been home for a long time, but you hear water running. You go upstairs and find a vigorously draining bathtub, which has only an inch of water in it. Now, it is possible that you showed up just as the last bit of water was draining out—but isn't it more likely that the tap has been left running? The lower the water level in the tub, and the faster it is draining, the shorter the lifetime of water in the tub and the greater the likelihood of a steady state, with a running tap.

This is the logic that led to the suggestion of a steady state of water on Venus. The low water level in the "tub" of the Venusian atmosphere, combined with the rate at which it seems to be draining, suggests that a tap is running somewhere. But what is the tap? What might be the source supplying water to Venus?

One possibility is volcanoes. The gas that hisses and burps from volcanoes on Earth is largely water vapor. As we saw above, the *Pioneer* orbiter observed variations in sulfur dioxide that some regarded as signs of active volcanoes. The topo map made by the radar altimeter on the orbiter showed some good candidates for volcanic landforms, but the map was too crude to tell whether they were really volcanoes and whether they were active. At the time of *Pioneer*, we knew enough to say that there probably is a volcanic water tap on Venus, but we could not determine whether it was wide open, running at a slight trickle, or stopped long ago.

We know of another source of water: Venus, like every other object in the solar system, is continually being pelted with snowballs, small ones fairly frequently, and very large ones more rarely. I am talking about the thin but massive horde of icy comets surrounding the distant outskirts of our planetary system. Occasionally, disturbed by the gravitational tug of a passing star or some other provocation, one of these frigid leftovers from planetary creation takes the plunge toward the vicinity of the Sun and boldly streaks through the inner solar system. As a comet approaches the Sun, its outer layer heats up and the ices begin to evaporate,* forming a massive tail of gas and dust streaking millions of miles in the antisunward direction. It's this huge, bright tail that makes comets (which in their normal frozen state are tiny, quite dark, and difficult to spot) such wonderful objects of telescopic, and often naked-eye, delight.

In 1986 the most famous comet of all, Halley's comet, made one of its close passages to the Sun, growing and showing off a new tail, as it does every seventy-six years. A lot had happened on Earth since Halley's last appearance, in 1910. This time it was met by an international fleet of spacecraft, which made close-up observations and directly sampled the cloud of gas and dust streaming away from the comet's frozen nucleus. These observations confirmed the "dirty snowball" model of comets. They are made roughly half of water ice, with smaller quantities of other hydrogen-containing ices. When these objects strike Venus (or any planet), which they certainly do on occasion, they are completely vaporized and their water becomes the planet's water. Even if they somehow survived the intense heat of the initial impact, these frozen space invaders would very closely approximate the proverbial snowball in hell, and would last about as long.

What a bizarre life these hydrogen atoms lead: sequestered in cold storage for billions of years in cometary ice, liberated in a violent impact on Venus with an energy of hundreds of millions of megatons, drifting for millions of years in Venus's torrid atmosphere in a variety of chemical combinations, diffusing into the upper atmosphere (perhaps doing some time in the sulfuric acid clouds on the way up) only to be flung back out into interplanetary space.

How much water do comets contribute over time? We don't know exactly. It depends on the total number of comets striking Venus over time, and the total mass of water frozen within them. We can make rough estimates using information from the number of craters on the Moon, telescopic observations of comets, and orbital calculations. Interestingly, the rate at which water is escaping from Venus now, as best we can determine, falls right in the middle of the range of predicted water infall rates. It is quite possible, based on our knowledge to

---

*To be more accurate, the ices don't actually evaporate, they *sublimate.* In the near vacuum of space, liquids (which require pressure to keep from instantly evaporating) never form, and as the ices heat up they turn directly to vapor.

date, that most of the water we find on Venus today comes not from the remains of an ancient ocean but from a continuing rain of dirty snowballs from the fringes of the solar system. In fact, in this scenario (a steady state with a cometary water source) the rate of hydrogen escaping from the atmosphere should equal the average source from comet impacts over the last couple hundred million years. If we could be sure that there were no significant water source from volcanoes, this could reveal to us the rate of comet impacts over this time interval. We could use the whole atmosphere of Venus as a giant comet detector!

The possibility that comets may be a major source of water on Venus introduces another wrinkle into the problem of trying to reconstruct the past from present observations. The size distribution of comets is top-heavy. If you took one hundred of them at random, most of the mass would usually be in the few biggest ones. So if you got rid of, say, the ninety smallest ones, the mass of the group would not be greatly diminished. This means that most of the water is contributed by the very occasional random impact of a very large comet. If this is the case, if this tap runs in occasional enormous spurts rather than a steady stream, then the water abundance might undergo large random variations,* rising with the impact of the largest comets and falling steadily in the long intervals between such collisions. But then how can we assume that the current water abundance is typical of long-term history? If there have been any very large comet impacts in the last couple hundred million years, then this assumption may not be valid (see Figure 3.11).

There certainly could have been such an impact: A huge one on Earth 65 million years ago wiped out 80 percent of all species then living, including those paleontological celebrities, the dinosaurs. This event paved the way for us mammals to walk the Earth without fear of being snacked on by giant reptiles. That comet, according to some estimates, might have contained more water than exists on all of Venus today. A similar event on Venus could have thrown both the water and D/H ratio *currently observed* out of whack, thwarting our efforts to extrapolate these quantities back through the ages.

---

*In the scientific literature we call these "stochastic variations." *Stochastic* is a word which means "random," which we use to make sure that no one can understand our papers.

3.11 Not a Near-Miss: *This photograph of Halley's comet was taken at Lowell Observatory in 1910. The other bright object in the frame is Venus, but the comet did not come close to the planet. However, comets do strike Venus, making craters and affecting its atmosphere over time.*

Nature sometimes sends us a dramatic reminder of this water source: just last night, a friend and I drove to the outskirts of town, where the lights stop and the sky is reasonably dark, to gawk at comet Hyakutake. For my whole life I have hoped for a comet like this to come. What a gorgeous sight—this evaporating icy intruder with its huge, faint tail, blown back across the night by the solar wind.* This comet, discovered only two months before it passed within 10 million miles of Earth on March 25, 1996, probably contains about 10 thousand billion (10^{14}) pounds of frozen water. If it were to hit Venus, it would increase the amount of water there by 10 percent.

If water on Venus is in steady state, we can still use the D/H ratio to infer the past water abundance, but we need to know how fast water is cycling through the atmosphere, from tap to drain, to determine the history. What is the actual escape rate of hydrogen at present? How

---

*Did you catch it, too? I hope so, but if not it will be back again in sixteen thousand years. And there will be other good comets. One called Hale-Bopp, coming in spring '97, may be spectacular. But comets are a bit like basketball teams. We can try to predict how the season will go, but sometimes they fizzle, or surprise us with a dazzling performance.

often are the frisky little hydrogens managing to get free of Venus's huge gravitational trap? Believe it or not, the attempt to answer this question has generated considerable debate. Why should this be controversial? Why can't we just send up a spacecraft like *Pioneer* to measure the flow? Because the escape rate from a planet is impossible to observe and difficult to estimate.

In the time it is taking me to type this sentence about 10^{27}, or a billion billion billion, atoms of hydrogen have cut loose from Venus to wander the solar system unchained from planetary gravity.* In the time it took me to stop and calculate the numbers to fill in the previous sentence and this one, another few times 10^{27} more of them got away. In one month this would add up to enough hydrogen to make an Olympic swimming pool's worth of water. The current escape rate seems to be about 10 million hydrogen atoms per second from each square centimeter of an imaginary sphere surrounding the planet's upper atmosphere, give or take a factor of several. That "factor of several" is the problem. It could be many times higher or many times lower. If we could completely surround Venus with a giant net, actually manifest that imaginary sphere, and count up the hydrogen atoms caught in it over a period of time, then we would know. But we can't, so instead we measure the densities and velocities of the particles involved at various places around the planet, fill in the blanks with reasonable models, and come up with the best number we can.

The overall rate at which H is escaping has proven to be a volatile quantity in the scientific literature. Individuals may seem pretty sure that they've got it right at any given time, but the numbers have bounced up and down over the last decade like airfares or the price of artichokes. Likewise, we are not sure how well the ban against deuterium escape (due to its greater mass) is enforced. We know that some D escapes as well as H, and the estimates for this ratio, crucial for the evolutionary models, have also changed dramatically, as the theory of atmospheric evolution itself evolves.† A detached observer (which I am not) might appreciate the great effort that scientists have put into this

---

*Any Avogadro fans out there can easily calculate in your heads that this is about two kilos of hydrogen. There is sufficient information in this paragraph to calculate my typing speed. This could get me a job someday.

†The theory is evolving much faster, but the atmosphere had a 4.5-billion-year head start.

modeling while also wondering whether their confidence in the latest numbers is fully justified. Yet these are the numbers we need to understand the history of water on Venus.

The uncertainties that obscure our view of the past history of Venus's atmosphere remain huge. The picture on the screen is lousy. Why do we bother to look, then? Because the screen is a time machine, and it is irresistible. How could you not look at a time machine? The D/H ratio is certainly telling us something about the past atmosphere. And as we chip away at the sources of uncertainty, and add other constraints from completely different areas of investigation, the number of possible pasts diminishes and the picture slowly clarifies.

Evolutionary modeling is difficult, but it allows us to connect currently observed quantities with various reconstructions of a planet's past. We cannot yet say which possible history is the "true" one. But we can rule some out, and we can make reasoned arguments for certain scenarios over others. This gives us something to debate and provides a basis for designing new experiments. It fuels and guides our imagination and intuition about how planets, including the one you are on right now, may have changed their atmospheres and climates, and how they may change in the future.

TRACES OF ANCIENT NOBILITY

Another important goal of the *Pioneer Venus* and *Venera* missions, also important for evolutionary modeling, was to try to pin down the amounts of the noble gases such as argon and neon. These atoms, with complete electron shells, are either too proud, aloof, beatific, or boring to interact with their atomic cousins, to bother with chemistry. Choose your own metaphor. They don't react. Since they do not participate in chemical reactions, their amounts in the atmosphere cannot have been changed by most factors that affect other gases. Their relative abundances probably reflect processes that date back to the formation of the planet. We hope that by studying noble gases we can find clues left over from planet formation. Differences in noble gas abundances between the planets may reveal differences in their formation and earliest histories.

The mass spectrometers on the *Pioneer* and *Venera* probes found very large amounts of argon and neon in the atmosphere— fifty to one hundred times the amount on Earth. This surprising discovery got the theorists buzzing.

Some researchers noted that these gases are more abundant in the Sun than in most planetary atmospheres, and thus were presumably present in large quantities in the "solar nebula," the gaseous cloud surrounding the young Sun during the process of planetary accumulation. Perhaps Venus somehow acquired much of her early atmosphere directly from the solar nebula. Earth and Mars, which lack all this argon and neon, may have completely shed their "primary," primitive, solar-type atmospheres early on, and evolved completely new atmospheres through geological and (at least in the case of Earth) biological activity. On a superficial level, this observation supported (and led to a brief revival of) the old notion that Venus's atmosphere was in a more "primitive" state than Earth's, trapped in the past.

The idea does not hold up under closer scrutiny. If Venus possessed such a primitive atmosphere, derived from the solar nebula, we would expect all the other noble gases to be similarly enriched on Venus, compared with Earth, in a pattern reflecting the composition of the Sun. They're not. *Pioneer Venus*, and the *Venera* probes, found that krypton and xenon are slightly enriched, but by a much smaller amount than argon and neon. More recently, scientists have proposed other ideas to explain the excess argon and neon, including the possibility that these gases arrived frozen within the ices of an early rain of comets or possibly even one giant comet. The alien pattern of noble gases does suggest that the origin or early history of Venus is different from Earth in some significant way. But no one has yet come up with an explanation that has a strong enough ring of truth to be widely embraced among comparative planetologists. This remains an area of lively debate and no small amount of confusion.

The *Pioneer* and *Venera* mass spectrometers measured not only the relative amounts of the noble gases but also their *isotopic compositions*. Just as hydrogen comes in more than one flavor, with one or two added neutrons in the nucleus making the heavier variations deuterium and tritium, so it is with the noble gases. Each one has different versions with slightly different masses due to different numbers of neutrons. The different mass versions of the same element are called isotopes. The particular mixture of isotopes within the inventory of an element contains information about the sources of this element on the planet in question.

Here more surprises lurked. We were especially puzzled by the isotopic composition of argon. Here on Earth, the most common kind is argon-40,* which is *radiogenic,* produced by the radioactive decay of potassium. The argon thus created in Earth's interior eventually winds up in volcanic gas. Its amount in the air is believed to reflect both the terrestrial concentration of potassium (a common element in rock-forming minerals) and the total amount of gas that has been added to the atmosphere by volcanoes.

In Earth's atmosphere we find four hundred times as much argon-40 as argon-36, its nonradiogenic cousin that is presumably left over from planet formation. Thus the ratio of argon-40 to 36 is four hundred. This high relative abundance of argon-40 results from Earth's long history of volcanic activity. The ratio we measured on Venus was four hundred times lower, roughly equal to one. Argon-40 and 36 are present on Venus in equal quantities! Despite its extremely thick atmosphere and huge total amount of argon, Venus has much less radiogenic argon-40 than Earth.

Only two explanations suggested themselves, neither very satisfying. Either Venus formed originally with much less potassium than Earth, or volcanic activity has been greatly inhibited on Venus as compared with Earth.† Either possibility presents a huge challenge to our ideas of comparative planetology. We expect such nearby planets, virtually the same size and density, to have been formed from similar materials, and a relatively common mineral like potassium should be roughly equally distributed between them. And we expect that, especially given their similar size, they should have had roughly equal amounts of geological activity. But this implies that their atmospheres should have had roughly the same amount of gas added by volcanoes. Something's not right. We will return to this topic below. As we dig deeper into the solar system, sifting for clues, one thing we keep learning is that in planetary exploration, expectation and discovery often have little correlation.

---

*The number refers to the atomic mass, or the combined number of protons and neutrons in a nucleus of this isotope. Argon-40 has four more neutrons than argon-36.

†The rate of radioactive decay is the same everywhere, so a given amount of potassium should produce the same amount of argon-40 on any planet.

AN OCEAN IN THE SKY

Another major area of investigation for *Pioneer* was the clouds. For centuries, ever since their presence was cleverly deduced by early telescopic observers, we have wanted to know the composition, extent, and structure of the global cloud layer that dominates the appearance of Venus. By the early 1970s, astute Earth-based observers had figured out that sulfuric acid was the main cloud-forming substance, at least in the uppermost layer visible from Earth. Several early fly-by missions and Earth-based telescopic observations had been used to characterize the properties of the very top of the cloud deck, and the thin overlying hazes, visible from above. But what of the layers invisible from the outside? How deep into the atmosphere do the clouds go? How does their thickness change with depth? Are they composed of more than one distinct layer? Are all the layers made out of the same kind of droplets?

Several probes prior to *Pioneer* had taken a stab at answering these questions. *Venera 8*, in 1972, had measured the disappearing sunlight during its descent and helped to define the approximate altitude of the cloud base. Beginning in 1975 with *Venera 9*, each entry probe had carried an instrument called a "backscattering nephelometer" to take a detailed profile of cloud thickness during descent.

You know how, when you are driving down a dark road at night with your brights on and you hit a patch of fog, the light scattering back into your eyes from the fog makes it harder to see? The thicker the fog, the worse this problem is, sometimes forcing you to turn off your brights or, heaven forbid, slow down a little. The intensity of the light coming back at you, the "backscattered" light, is a measure of the thickness of the fog. This is exactly how the backscattering nephelometers on the entry probes worked. They shone a light into the darkness as they plummeted and recorded the changing intensity of the light bouncing back to them, which reflected (literally) the changing thickness of the clouds.

The nephelometer traces from the different probes, each a profile of changing cloud thickness with altitude, are shown in Figure 3.12. Each one is a slice through the Venusian clouds at a more or less random place and time. As you can see, they are all different. The answer to the question of vertical cloud structure was different each time we asked it. Even though from the outside the clouds appear uniform and homogeneous, when you drop down through them at different places

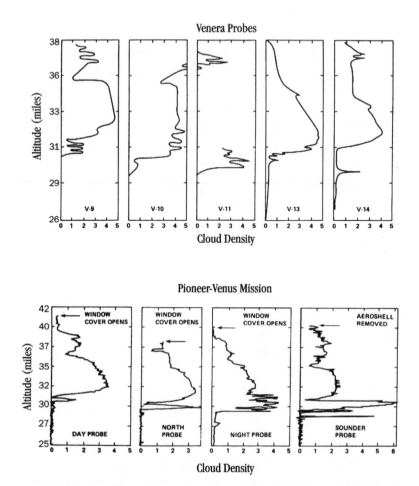

3.12 Cloud Profiles: *These are profiles of cloud thickness with altitude made during descent by five different Venera probes and the four Pioneer Venus probes. They are all different, which tells us that the clouds are always changing.*

and times, they turn out to be quite complex and constantly changing. Just glancing at the variety in these profiles, we can conclude that there is no such thing as a "typical" cloud structure on Venus.

Nonetheless, from these experiments and others, we have deduced some more or less invariant (unchanging) characteristics of the clouds. They are usually divided into three layers with relatively clear air in between them (see Figure 3.13). There are three distinct kinds of cloud droplets, of different sizes: minuscule, tiny, and large.

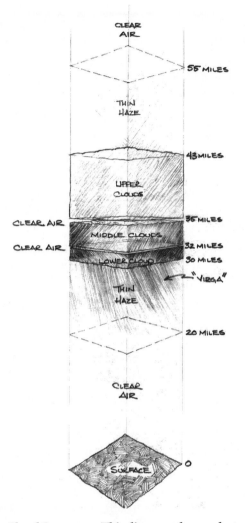

3.13 Average Cloud Structure: *This diagram shows what we believe to be a typical structure in the volatile clouds of Venus. (Carter Emmart)*

We call these "modes" 1, 2, and 3, and we argue about whether the different sizes imply that they are made of different materials.*

*Mode 1 droplets are .00002 inches across, roughly the size of smoke particles. Mode 2 droplets average .0001 inches, similar to drops in a light fog on Earth. Modes 1 and 2 are present in all cloud layers. Mode 3 particles are much larger, with diameters up to more than .001 inch. These are more similar in size to cloud droplets on Earth, and they dominate the lowest cloud layer.

Against the background of this more-or-less constant overall structure, the details of the clouds at any particular place and time can vary a lot. This is not surprising given the fact that the clouds exist in an altitude region where winds are fierce. The entry probes also found several layers of very high turbulence within the main cloud deck—the same kind of turbulence that lights up the "fasten seat belt" signs on commercial jet flights and makes the cabin begin bouncing around mercilessly every time you are about to dig into your delicious airline meal. This kind of atmospheric overturning on Venus and Earth (or anywhere else) occurs when there is an unstable distribution of atmospheric temperatures, one that gets too hot too fast with declining altitude. The air readjusts with rising and falling (hot and cold) currents, jostling aircraft, spacecraft, clouds, and sometimes your stomach.

If anything, it's surprising that some characteristics of the cloud deck seem to be even sort of constant, given this turbulent, dynamic environment. The clouds are constantly remaking and refreshing themselves against the vigorous forces that would otherwise completely shred them.

Although sulfuric acid seems to dominate the composition of the clouds, the actual chemical composition of the cloud deck may be considerably more varied and complex. That separate cloud droplet populations with different sizes (the modes) exist is harder to explain if they are all made of pure sulfuric acid.

Several different experimental results have hinted that the large "mode 3" particles may be all or partly made of some other substance. Some *Pioneer* results seemed to indicate that these might be non-spherical in shape. They could be solid crystals. If so, they must be made of some stuff with a higher freezing temperature than H_2SO_4. The only experiment that actually collected cloud particles and analyzed their chemical composition was carried by the Soviet *Venera 12* mission. It found that the most common element in the clouds was not sulfur but chlorine, the stuff of swimming pools, table salt, and stomach acid! The clouds held about twenty times as much chlorine as sulfur. These results are so difficult to reconcile with other measurements that American researchers have tended to ignore or discount them, although no one has explained why they should be in error. Other experiments in which we've probed the density structure of the clouds with radio waves suggest that the lower cloud droplets may be

made of something else and *coated* with sulfuric acid. We have to conclude that the composition of the large mode 3 droplets, and the question of whether they are hiding something unknown inside
themselves, is still uncertain. Yet most of the mass of the clouds is in
the form of mode 3 droplets, and most light is absorbed by them. It is
possible, given current knowledge (or lack thereof), that most of the
planetwide cloud deck is made out of some mystery substance that has
not been identified.

So far I have described the structure of the clouds, the dynamic
motion of the atmosphere in the cloud-forming region, and the light
absorption there as if all three of these were completely separate, independent entities. In fact, the cloud structure, winds, and radiation
form a very complex system of mutual interactions. The clouds block
and filter sunlight, determining where solar radiation gets absorbed.
The spotty deposition of solar radiation leads to differential heating of
the atmosphere, which causes turbulence and winds. The turbulence
and winds move around the cloud material and determine where it
forms and vanishes (which affects where sunlight gets absorbed ...).
You can tell that this whole system involves many complex feedback
loops; or, as we say in the industry, the problems are "coupled." Models
that aim to do a good job of describing any of these phenomena
(clouds, winds, or radiation) must really describe all of them, as well as
the feedbacks. This is no easy task, but who ever said that understanding Venus was going to be easy?

We've looked at the Venusian clouds from all sides now, from
above, below, inside, and out. This might be a good time to examine
how they stack up alongside the clouds of Earth. What we find is that
they are clouds, in the terrestrial sense, in name only. Like our clouds,
they are composed of condensed droplets suspended in the air above
the planet, but that is where the resemblance ends.

There is the obvious difference that Earth's clouds are made
mostly of water and water ice, with some trace impurities, whereas
clouds on Venus are made mostly of sulfuric acid, probably with other
stuff mixed in. Unfortunately, Earth too now has clouds with high con-
centrations of sulfuric acid downwind from certain urban areas. So we
are doing our best to simulate the clouds of Venus, with dire environmental consequences, including the acidification of the great lakes of
North America and the dying-off of forests in Europe. We will never

live to see the day that the clouds of Earth are acidified to the extent of those on Venus, hopefully because they never will be.

If you look up from the surface or down from above, the clouds of Venus are opaque everywhere. You can never see the surface from orbit (at least at visible wavelengths) and you could never see the Sun or stars from the surface of Venus. But the clouds are actually quite diffuse and composed of extremely tiny droplets. The average size of cloud droplets on Earth is similar to the size of the largest mode 3 droplets found on Venus. The *average* density of Earth clouds is more than ten times as high as that on Venus, and the maximum density is one hundred times higher. Visibility in the clouds of Venus is much greater than in most kinds of clouds on Earth. In an airplane on Earth traveling through thick clouds you can't even see the wing tips of the plane you are in. In the clouds of Venus you could see for miles. So if you were ballooning there you would feel as though you were in more of a thin haze or fog than a cloud. But it's a fog without end, which keeps going horizontally around the whole planet and vertically for more than ten miles. It is the thirteen-mile height of the Venusian clouds, not great density at any given height, that makes them so opaque.

In many ways, the clouds of Venus are their own entity, not like terrestrial clouds at all. In fact, in some ways the Venusian clouds are really more like an ocean. Venus doesn't have an ocean. It may, or may not ever, have had one. Today, at least, it's much too hot for any substance we are used to thinking of as a liquid (like water or sulfuric acid) to condense on the surface. But maybe, to avoid the heat, Venus just chooses to form her ocean thirty miles above the ground, where it's cool enough for condensation.

The oceans of Earth are a vast reservoir of condensed liquid, nearly global in extent, which circulate around the planet, controlling the temperature and serving as an incubator for complex chemistry among the constituents dissolved there. The clouds of Venus also are a globally extensive reservoir of condensed materials. They also control the temperature of the atmosphere, not because of a large mass (like Earth's oceans) but through their role as gatekeepers, determining which radiation is let in and out of the subcloud atmosphere. On Earth, much of the Sun's light is absorbed at the surface of the ocean. On Venus, the cloud tops play this role. Although not dense, they are everywhere much deeper than the deepest of Earth's oceans, and they circulate around their world in great globe-spanning currents. As for exotic chemistry, who knows? We do know about some interesting sulfur chemistry, probably including the unknown ultraviolet absorber that resides in the cloud tops. There are hints of other unknown substances hiding in the cloud droplets. The environment is much more stable than that of the more volatile terrestrial clouds; particles in the upper clouds probably last for months, rather than for days as on Earth. Our clouds are a patchy presence riding the currents of weather, appearing and disappearing with the wind. On Venus the clouds are a constant global fixture always filling an altitude region more than thirteen miles high. This plus the constant ultraviolet sunshine creates a more stable and energetic environment for chemistry, which, if not exactly like that of an ocean, may be somewhere between this and what we normally think of as a cloud.

THINK GLOBALLY, PROBE LOCALLY

During that hectic Venusian December, the *Pioneer* and *Venera* entry probes gave us a much more complete portrait of the planet's atmosphere than we had had before and allowed the creation of a standard model, published in the *Venus International Reference Atmosphere*. But the discerning reader may wonder, "What if December 1978 was not a typical time on Venus? Wouldn't our 'standard model' be skewed by events or conditions peculiar to that time?" Good question.

In planetary science we are often forced to make do with such snapshots. We cannot always choose when to do our fly-bys or landings. Sometimes the timing is determined by celestial dynamics (when the planets are lined up just right for us to make the journey), and sometimes by the dynamics of congressional politics. But planets change, sometimes on short timescales.

Much of our understanding of the atmosphere of Mars comes from the incredibly successful *Viking* mission of 1976 that brought two landers and two orbiters to the red planet. These data informed our standard models of the Martian atmosphere for two decades. But *Viking* went to Mars, it now seems, at an atypical time, when the atmosphere was dustier and warmer than usual. Recent observations with the Hubble Space Telescope have taught us that the air on Mars is often cooler and clearer than we had thought.

Any aliens surveying Earth for its investment potential, had they arrived and begun observing in 1991, would have found an unusually warm stratosphere with a thin sulfuric acid aerosol cloud, and a lower atmosphere a bit cooler than usual. If they were patient enough to watch for a year or more, they would have seen these perturbations (caused by the eruption of Mount Pinatubo in the Philippines) begin to subside. In the last few years, we have watched through our telescopes as huge storms, bigger than Earth, have suddenly come and gone on Saturn after decades of relative calm. One of the strangest places in the whole solar system is Jupiter's hypervolcanic Moon Io, which seems to be a little different every time we look at it. In the midst of all this change, we have to try to sort out the invariants—the persistent traits—from the more transient aspects. This is more challenging in those cases when we get only one good look.

We have to worry about two kinds of variability, really: *temporal* variability (things changing over time), and *spatial* variability, as exemplified by the problems of trying to define a "typical" cloud structure, or a "typical" place to land and photograph a planet. These would be less of a problem if we had unlimited resources to thoroughly explore the solar system. With enough planetary missions we could observe each planet over time, noting how they change. And we could probe enough places in the planets' atmospheres, and land in enough places on their surfaces, to build up an understanding of the spatial variations. Dream on. Instead we have to be clever, and sometimes lucky.

An extreme example of this kind of problem occurred in December 1995, when the *Galileo* spacecraft reached Jupiter and sent a probe into its atmosphere. *Galileo* had a long, troubled journey to Jupiter that included, for the gravitational slingshot effect, one close fly-by of Venus and two of Earth. During its travels, the umbrella-shaped main antenna for communications with Earth failed to open, and the onboard tape recorder, crucial for storing information for much slower playback using the backup antenna, also had major problems. When *Galileo* finally made it to Jupiter, it sent its probe plunging into the atmosphere and clouds of the most giant planet—a probe that owed much of its successful design to experience gained from the *Pioneer* and *Venera* probes at Venus. It was a rare opportunity indeed—the first probe ever to directly examine the atmosphere and clouds of Jupiter, and the last one for the foreseeable future. There are no plans, as far as I know, to send a follow-up.

Given this unique opportunity, it would have been nice to be able to examine a "typical" or "average" spot (to the extent that such concepts are meaningful on such a turbulent, volatile world). Instead, nearly the opposite happened. The initial results were quite puzzling. Nephelometers and mass spectrometers (and other instruments similar to those on the *Pioneer* probes) measured the cloud properties and, incredibly, the observations showed no clouds at all. Then ground-based photographs revealed why: The *Galileo* probe fell through a hole in the clouds of Jupiter! The probe appears to have entered, by pure bad luck, into a very rare spot on the disk of Jupiter where the clouds were so thin as to be almost invisible. The chance of falling into a spot like this, given the almost complete cloud coverage on Jupiter, is about one in one hundred. This is akin to an alien Earth probe landing in Death Valley to measure Earth's water. You pays your money and you takes your chances.

So, was our one-in-a-million chance to learn as much as we can about the solar system's big boss planet compromised? Yes! The *Galileo* scientists, beleaguered by years of delays and technical problems, seemed like they didn't really want to admit this at the initial press conference called to announce the probe results, but the bizarre point of entry does raise doubts about the general applicability of some of the findings. Does this mean the probe results are useless? No! They are tremendously valuable, but the incident reminds us of the hazards of trying to understand a changing, complex, and turbulent world from a probe that studies one place at one time.

We have had many opportunities to look at Venus, but none as thorough, in terms of an in-depth look at the atmosphere, as the numerous simultaneous probes of *Pioneer* and *Veneras 11* and *12*. How was Venus atypical in 1978, and how has it changed? First, I should say that most likely it was atypical then in some ways that we haven't figured out yet, since we have not had as good a look since then. But we do know of several changes. As mentioned above, the SO_2 abundance above the clouds has been declining ever since then, possibly because of the fading effects of a giant volcanic eruption. Along with this, the amount of thin above-cloud haze has also been declining. The structure of the main cloud deck looks relatively constant viewed from the outside, but we now know that the clouds are patchy and highly variable in the turbulent lower layer, that which contains most of the mass. Had the probes fallen at slightly different places or times, our standard model would be different. This is one of the limitations that planetologists have to live with. But clever observers on Earth have devised new ways to monitor some of these changes.

NEW WINDOWS INTO THE DARK SIDE OF VENUS

Once in a while you get shown the light in the strangest of places if you look at it right.
—ROBERT HUNTER, "Scarlet Begonias"

Eleven years would pass after the launch of *Pioneer* before another American spacecraft would be sent to Venus—a craft called *Magellan,* which would finally and dramatically reveal the long hidden surface. But planetary scientists on Earth were not sitting idle. Several events during the 1980s helped to fill in the blanks in the picture created by *Pioneer* and to whet our appetites for the feast of images that, if all went as planned, *Magellan* would provide.

There was still a hugely unsatisfying order-of-magnitude uncertainty about water abundance in the lower atmosphere, and large question marks regarding the amounts of some other important gases remained. But with a little luck, a lot of craftiness, big telescopes, and supercomputers, we found a secret passageway into the lower atmosphere.

In 1984 the Australian astronomer David Allen needed to test a new infrared camera that he had developed for the giant telescope at the Anglo Australian Observatory in Coonabarabran, New South Wales. David was an extragalactic astronomer: much of his research dealt with objects far beyond even our own immense galaxy of 100 billion stars, let alone the tiny swarm of planets and other debris orbiting our own star.[*] It is a joke among planetary scientists, based on some not insignificant grains of truth, that such people regard the objects within our solar system (planets, asteroids, comets, moons and Pluto)[†] as mere annoyances, obstacles to be avoided for the much more important business of observing the rest of the cosmos. But David needed to test his camera, and for this purpose he wanted a bright, boring object that would be "featureless," without any detailed pattern of absorption,

---

[*] Alas, the past tense is necessary. David Allen passed away in 1994.

[†] As this book goes to press, scientists have still not resolved the furious debate over whether Pluto is or is not a planet. I consider this burning question far too important for me to offer an opinion.

in the near-infrared part of the spectrum. The nightside of Venus seemed the obvious choice for such a test, before he turned his new equipment on more interesting objects. He fitted his telescope with the new camera and pointed it at Venus, fully expecting to see the bland and predictable near-infrared signature of sulfuric acid at –45 degrees, the temperature of the Venusian cloud tops.

What he found instead was serendipity, the joy of accidental discovery. The spectrum was anything but bland. When he graphed brightness versus wavelength, he saw two huge spikes. The nightside of Venus was radiating obscene amounts of radiation in two near-infrared colors, narrow spectral intervals, at 1.7 and 2.3 microns (see Figure 3.14).* After a couple of false starts, he hit on the correct interpretation for this "anomalous" radiation: heat radiation was leaking from the lower atmosphere at those wavelengths. This came as a big surprise because we had previously assumed that the atmosphere was completely opaque at all near-infrared wavelengths, blocking radiation into space from the lower atmosphere. After all, CO_2 is a damn good infrared absorber and we had thought the small holes in the CO_2 spectrum were effectively plugged by minor amounts of other gases. But it turns out that there are a couple of narrow intervals where the overlapping absorption patterns of CO_2 and water vapor (the other major absorber) do not fill in all the holes. The windows are—you guessed it—at 1.7 and 2.3 microns.

These infrared leaks in the otherwise hermetic Venusian atmosphere provided much more than an interesting curiosity, a new phenomenon for us to explain. There was a further, exciting implication. For decades we had pointed telescopes with spectrometers at Venus and seen only clouds and the spectral signature of the thin atmosphere above them. We had given up trying to see beneath the clouds in the infrared. The discovery of these "window regions" in the nightside spectrum meant that, for the first time, we could sense the lower atmosphere directly from telescopes on Earth. We had found an open window, through which we could glimpse the strange vapors lurking deep

*A micron is a millionth of a meter in length. The wavelengths of visible radiation go from .4 to .7 microns. The anomalous radiation from Venus being discussed here has wavelengths just a few times longer than visible light, so it is in the part of the electromagnetic spectrum we call the "near infrared," because it is just beyond the red edge of the visible.

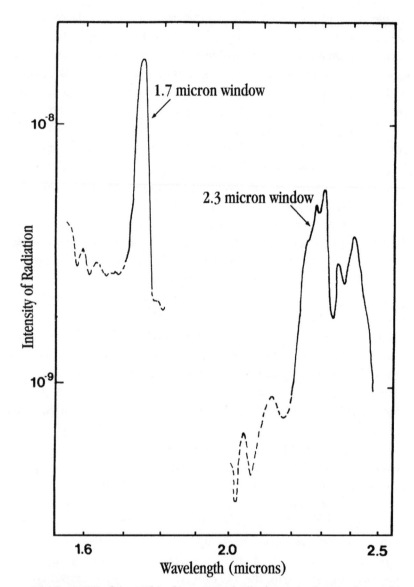

3.14 Open Windows: *This shows David Allen's original spectrum with which he discovered the near-infrared windows in the nightside of Venus. The excess of radiation at 1.7 and 2.3 microns is leaking from the hot lower atmosphere. (D. Allen)*

in the atmosphere of Venus. We knew that the intricate details of the shape of these windows contained precise information about the quantities of trace gases in the deep atmosphere. It seemed possible, at least in theory, to take much more detailed spectra of the light escaping from these window regions, and use them to directly study the lower atmosphere, without having to get a spacecraft to fly through NASA, Congress, and across interplanetary space. Even after several successful spacecraft probes, the gas composition was far from clear. Could we now probe the sweltering depths of Venus from the (relative) comfort of Earth's icy mountaintops?

The actual realization of this dream took years of hard work on two fronts: observation and theory. Several of our world's best observers, wishing to flesh out the finer contours of the newly discovered windows, took new and improved near-infrared spectrometers to some of our world's best telescopes—in Hawaii, in Australia, and even in the lower stratosphere mounted in a jet aircraft. Each observing run revealed some precious new wrinkle in the spectral windows. Figure 3.15 shows one of these detailed spectra. Note the increased detail over David Allen's "discovery spectrum," in Figure 3.14.

The theorists had some heavy homework to do before we could make use of this mother lode. The main problem may come as a surprise: We needed to know the near-infrared spectrum of CO_2. We needed this to find the needles in the infrared haystack, to tell which of the "squiggles" in the spectrum were due to the huge background of CO_2 and which were due to the trace gases we were trying to pin down.

Now, you may be wondering, how could it possibly be that we didn't have a complete map of the infrared spectrum of a common gas whose infrared absorption properties are key for climate studies of Earth? Yes, it's true that CO_2, most notorious of all greenhouse gases, is one of the key ingredients in global change on Earth. Its study has been a matter of great urgency and the target of numerous research megabucks. Indeed, we have carefully mapped its spectrum, largely in work funded by the U.S. Department of Defense.* But it turns out that these data are not very useful for understanding the infrared spectrum of CO_2 on Venus, where it behaves differently because of the hot environment. On a molecular level, the heat means that many more molecules are in the "excited" states resulting from energetic collisions, and they absorb wavelengths that are just ignored by the relatively calm terrestrial CO_2 molecules.

*This is one example of "defense" spending that clearly has contributed to our security, our defense against global warming.

This is something we have to watch out for in planetary science: the properties of common substances can be quite different on other planets. If you take a familiar beast to a place as strange (to us) as Venus and let it loose, it may behave in ways that will surprise you. We may think we have a complete understanding of something as common as CO_2 gas, but when we needed to know the spectrum of CO_2 on Venus, in detail, we found that we knew very little about how it absorbed near-infrared light at Venusian temperatures.

Some very clever theorists using quantum mechanics and many hours of supercomputer time soon solved the problem, creating a new "high-temperature spectral database" of CO_2. This intricate map included literally millions of hitherto unknown spectral features.

Armed with this new spectral map for high-temperature CO_2, we could finally, with confidence, simulate the observations on a super-computer. We generated synthetic spectra to compare with the real ones gathered with telescopes and spectrometers. We knew that the features not due to CO_2 were caused by some of the trace molecules we were seeking. We had carefully mapped every piece of straw in the hay-stack, and the needles started to jump out at us. This technique has proven to be a fantastically rich source of data on the lower atmo-sphere. We have found, in these windows, the spectral signature of water, sulfur dioxide, carbon monoxide, and several other gases we have long wondered about. We vary the amounts of these gases in our computer models, compare the "virtual" spectra we generate with the real observed ones, and nail down the amounts of these important trace gases by seeking the best match. The greatly improved under-standing of the lower atmosphere is allowing us to create much more realistic chemical, evolutionary, and climate models. Figure 3.15 is an example of a high-resolution spectrum of the 1.7 micron window, alongside synthetic spectra made with three different water abun-dances. That the best fit is for thirty parts per million is easy to see. This is how we finally figured out just how dry it is down there.

These results showed conclusively that Venus was even drier than many of us had thought. We can now place the water abundance securely at the dry end of the previous order-of-magnitude uncer-tainty. This means that the lifetime of water in the atmosphere is even shorter, increasing the likelihood that the water we see now comes

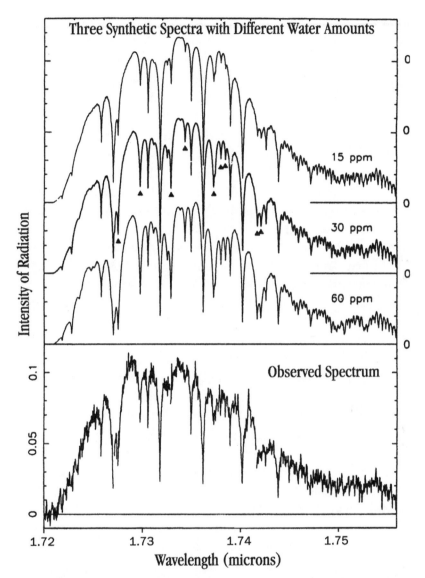

3.15 Needles in the Infrared Haystack: *The curve on the bottom is a very detailed spectrum of the 1.7 micron window seen in the previous figure. The three curves on top are synthetic spectra made assuming an amount of water in the atmosphere of (from top to bottom) fifteen, thirty, and sixty parts per million. The triangles show absorption features made by water. Note how the shape of these features fits the observed spectrum best for a synthetic spectrum made with thirty parts per million water.*

from a recent source of comets or volcanoes, and is not the last traces of a disappearing ocean. (The level in the tub is really low, but it is draining vigorously. The tap must be running or it would all be gone.) This also means that our climate models of Venus, which work by including water as a trace greenhouse gas (like CO_2 is on Earth), all need to be redone.

Another finding from this work concerns the *gradient* of water in the lower atmosphere. (Remember this? It gets drier as you go deeper.) Observations in the window regions showed us that the gradient, which had caused us such headaches and intellectual contortions in attempts to explain it, almost surely did not exist. Water is at thirty parts per million all the way down. The water gradient seems to have been an artifact of our earlier experiments; all those clever explanations I described earlier, all the great arguments we had over this, instantly evaporated like a puddle of water on a hot Venusian night. It was fun while it lasted. This kind of thing is inevitable when exploring worlds.

The discovery of these windows created an additional opportunity. It allowed us to make images of the long-hidden lower clouds. David Allen and other observers had noticed that the bright radiation is leaking from small spots on the planet, and these move around. They correctly inferred that patchiness in the clouds creates this effect. In spots where the clouds are thinner, more radiation can escape. Since the clouds are obviously not patchy in the visible upper layers, the holes must be in the lower layers. This is consistent with the results of *Pioneer* and other cloud probes that had hinted at a lot of variability in the deeper, turbulent parts of the clouds that we cannot see from orbit or from telescopes on Earth.

Several observers started to monitor the motions of these bright spots, looking for a rotation rate, just as centuries before many had mapped chimerical spots in vain efforts to find a rotation rate of the elusive solid surface. But these bright infrared features were very real, and we soon found that they move in the same direction as the four-day superrotation. They are somewhat slower, however, showing rotation rates of five to seven Earth days. Since *Pioneer* had shown us how wind speeds vary with depth, this rotation rate told us the altitude of the features. The motion of these features matched the somewhat slower supperrotation wind speeds in the bottom layers of the clouds. A consistent story was forming. Sporadic clearings in the lower cloud deck allowed near-infrared light to escape into space, forming the bright spots with their rich spectra full of priceless information about the lower atmosphere. If we could take a really detailed snapshot of Venus in these same wavelengths, we would have a picture of the deep cloud structure over the whole nightside hemisphere at one time.

Remember, the problem with probes is that they show us only a slice from one place in the highly variable clouds. An extensive 3-D portrait, showing the large-scale structure, would be extremely valuable. Fortunately, in 1990, we had a spacecraft headed for Venus that was ideally equipped to take such an image. Just as we had gotten a free ride to Venus with *Mariner 10* en route to Mercury in 1974, once again we were able to freeload a Venus fly-by. This time it was *Galileo* that gave us a ride. The same craft that finally arrived at Jupiter and sent in a probe in 1995 (and is now studying the giant planet and its moons from orbit), had to fly by Venus in 1990 to receive a "gravity assist" toward the outer solar system.

Would you like to swing on a star? Would you settle for a planet? Just as *Mariner 10* had done sixteen years earlier, *Galileo* was to dive at Venus, grab hold of its mighty gravity, and swing itself onward, eventually to reach Jupiter. This ship is equipped with a fantastic set of instruments for studying the Jupiter system, and we were able to use most of them at Venus during the brief fly-by.

Galileo has an instrument called NIMS (near-infrared mapping spectrometer) that is perfect for taking detailed images at near-infrared wavelengths. NIMS can take a large number of simultaneous images all at different near-infrared wavelengths.* This is so well suited for studying the Venusian clouds that it's hard to believe it was developed to study Jupiter. On February 10, 1990, while *Magellan* was also en route to Venus, *Galileo* whipped close by the planet and was flung back away from the Sun. As the beleaguered, intrepid interplanetary wanderer closed toward Venus, NIMS was prepared to take a close infrared picture of the nighttime hemisphere.

But there was a problem: NIMS was too hot. Infrared detectors must be quite cold to work properly; on Earth we use liquid nitrogen to

---

*Or a large number of simultaneous spectra, all arrayed like an image. It depends on how you look at it. NIMS generates a 3-D data set, with two spatial dimensions and one spectral dimension.

cool our detectors at telescopes. In outer space it's not usually a problem, but sunlight reflecting off of some part of the spacecraft was apparently heating NIMS just enough to cause trouble. The overheating was only discovered shortly before the narrow window of time during which we could take this unique portrait. A command was sent to reorient the spacecraft and cool NIMS down. We anxiously watched the stream of numbers as the temperature dropped, with painful slowness, and the moment of truth approached.

NIMS just made it under the wire, cooling enough to produce images and spectra, instead of just garbage which we might have gotten had it been a few degrees hotter. A false color infrared image taken by NIMS as *Galileo* neared closest approach to Venus is shown in the color insert. Here you can see the deep structure of the clouds laid bare. This image is taken at 2.3 microns, one of the spectral "windows." You are looking at the nightside of Venus (which is black in visible light), seeing heat radiation leaking from the lower atmosphere. The structure you see is all in the lower part of the clouds, which we could never see before the windows were discovered. The bright spots are clearings where turbulent air currents are dissolving the cloud particles. Where the image is dark, the clouds are thick and radiation can't get through. You can see details, deep in the clouds, as small as thirty miles across.

NIMS took simultaneous images, just like this one, at seventeen different wavelengths. Each wavelength is sensitive to cloud structure at slightly different altitudes, and when analyzed together, they gave us a true 3-D picture of the clouds at the moment of the fly-by. It's interesting to compare the NIMS image in the color insert, with the cloud profiles made by the entry probes, in Figure 3.12. The image gives us a portrait of the clouds over a whole hemisphere, frozen at one moment. The probe profiles each give us a detailed look at one place, at different times. The whole data set is greater than the sum of the complementary parts. We now have more of a handle on cloud structure and variability than we did before the *Galileo* fly-by. This, in turn, has helped us with that thorny problem of trying to understand the coupled, interacting motions, physics, chemistry, and radiation in the clouds and surrounding atmosphere.

Now that we have this detailed portrait from NIMS, we know how to interpret the bright and dark infrared spots we see from telescopes on Earth. We can watch them move across the nightside and follow the changing weather in the previously hidden lower clouds of Venus.

Although the American spacecraft exploration of Venus took a long pause during the 1980s while ground-based observers worked their magic, the Soviet program didn't miss a beat during the first half of that decade. Well, they missed one, sitting out the 1980 launch window while they digested the feast of information from *Veneras 11* and *12*. But for the next three launch windows they were back with three successive pairs of new missions, all innovative, daring, and successful.

In 1982 *Veneras 13* and *14* landed and took the first color panoramas at the surface of Venus. These were the first surface pictures taken since *Venera 10* in 1975. (The cameras on *Veneras 11* and *12* had failed, reportedly because the lens caps became stuck on and melted over the cameras.)* A color picture from *Venera 13* is shown in the insert. Venus proved difficult to photograph in color. A "calibration strip" was included on the spacecraft so that its appearance in the photos could be adjusted to the known colors, ensuring accuracy. Unfortunately, in the environment of the Venusian surface the colors of the strip itself promptly changed.

So, what color *is* the surface of Venus? The answer is not as straightforward as you might think. It depends on what you mean by color, whether it is a property of a surface, or a sensory perception you would have if you were looking at it. Soviet and American scientists worked together to analyze spectral information from *Veneras 13* and *14*, partly to see if the surface is an oxidized (rusted) red like the surface of Mars. They found that, at least where the craft landed, the surface is dark gray, like basaltic rocks on Earth. It reflects short wavelength (blue) light about as strongly as it reflects longer wavelength (red) light.

But the light at the surface has a strong red tint to it. This is caused by the same phenomenon, known as Rayleigh scattering, that gives our sky here on Earth the blue we know and love. Atmospheric molecules scatter blue light much more strongly than red light, so on

---

*A Russian colleague related this anecdote: As the first (non)picture from *Venera 11* was received, in which nothing could be seen but melting lens cap, the camera designer exclaimed, "We seem to be in some dark, viscous liquid." A team member immediately replied, "Yes, sir, we are in deep shit!"

Earth the light that gets scattered out of the incoming beam of sunlight and lights up the sky is tinted blue.*

On Venus, however, any light that reaches the surface has come through the planetwide cloud cover and an atmosphere ninety times as thick as Earth's. Thus Rayleigh scattering effectively removes *all* the blue light, creating a planetwide red-light district. If you were standing on the surface of Venus (hopefully with your envirosuit in good working order), everything, including the surface rocks, would look red. (Of course, virtually the whole planet remains unphotographed, so for all we know there could be brilliant swirls of vermilion and gold just over the hill from *Venera 13* or *14*.)

These, the final two in the incredibly successful series of *Venera* landers, also managed to scoop up samples of surface material and analyze their composition with experiments inside the landers. This is no easy task if you consider the climate outside; you can't leave the door open for long. They found more evidence of basaltic sand and dust.

Venera 14 experienced one stroke of truly extraordinary bad luck. An arm of the lander was supposed to swing down and penetrate into the surface with a sharp pointed probe, measuring its properties. But it got stuck on something hard, and apparently artificial, manufactured by a technical civilization. Those damn lens caps again. This time it came off the camera smoothly, only to land in the one spot where it would block the surface probe. Soviet scientists calculated the odds of this happening as one in a thousand.

In 1983 two more Soviet craft, *Veneras 15* and *16*, entered Venus orbit carrying radar mappers. These craft successfully used cloud penetrating radar to image a large portion of the Northern Hemisphere. The quality was not nearly as good as we expected to get from *Magellan*, and the surface coverage was only partial. But the resulting maps gave us a lot to talk about. They showed surface features as small as one mile across, revealed Appalachian-style folded mountain belts, vast volcanic plains, large areas of heavily fractured crust, and a density of impact craters that suggested an intermediate age for the surface: older than Earth but younger than the Moon or Mars.

---

*What you see as the daytime sky is light that has been "scattered," bounced around by air molecules until it finds your eyes. On the Moon, with no air to scatter light, if you blocked out the Sun and let your eyes adjust to darkness, you could see the stars at high noon clearer than on the darkest night on Earth.

The Soviets sent two craft to greet Halley's comet in 1986, and on the way there each skimmed by Venus and delivered a package: *Vega-1* and *Vega-2*. Each *Vega* craft had a lander and an instrumented balloon. These were the first balloons to be sent to another planet, something we had fantasized about for decades. Each drifted around the planet for about two Earth days, carried westward for six thousand miles within the lower clouds by the rapid superrotating winds. An international team of scientists monitored their motions, which revealed a lot about turbulence and wind speeds.

The *Vegas* were the last Venus probes sent before the Soviet Union dissolved, in 1991, ending the cold war that had caused so much trouble on Earth* and that had spawned so much planetary exploration as interplanetary displays of international machismo. It was a grand finale. Hopefully in the future we will not need the prospect of "mutually assured destruction" to inspire us to explore the solar system. This was never the motivation of the actual explorers, only their sponsors. The explorers themselves are motivated most strongly by simple, insatiable curiosity. We may never infect our world leaders with this drive. If they had it, they would be scientists (or artists, historians, musicians, poets, architects, etc.) not politicians. However, there are other strong motivations. One of the lasting, and hopeful, legacies of cold-war madness is the huge amount of knowledge we have gained about the workings of worlds, including our own. If we are to survive here on Earth for a long time, long enough to thoroughly explore our planetary system and begin to explore the galaxy, we must of course learn to get along a bit better. But we also need to learn much more about how our planet works. This is no longer a luxury. We have unwittingly become planetary managers. Fortunately, we are getting a crash course in how planets work from the lessons of comparative planetology.

*And opening the gates for much new trouble.

4

chance or necessity?: sizing up the planets

I can hear the sizzle of newborn stars, and know anything of
meaning, of the fierce magic emerging here. I am witness to flexible
eternity, the evolving past, and I know we will live forever, as dust or
breath in the face of stars, in the shifting pattern of winds.

—JOY HARJO, *Secrets from the
Center of the World*

TERRESTRIAL PLANETS

For thirty-five years now, beginning in 1962 with the dispatch of *Mariner 2* to Venus, we have been exploring the solar system with spacecraft. These compact robotic agents of our curiosity have allowed us to examine our fellow travelers around the Sun in sufficient detail to finally see them for what they are: our sibling worlds. And as in any reunion with long-lost relatives, we find reassuring similarities and puzzling differences.

Earth has a lot in common with Mars and Venus. They are small rocky worlds with atmospheres and many familiar surface features that lend themselves to study by terrestrial analogy. Perhaps it is not surprising that we should share many similarities with these two, our closest planetary neighbors. After all, we share a common galactic heredity in the collapsing molecular cloud that became the Sun 4.5 billion years ago, spinning off the planets in the swirls of its afterbirth, and we've matured together under the steady incubation of the same slowly warming star.

We also find worlds of difference between Earth and our neighbors, and our young science is still seeking a deep theoretical understanding of many of these variations. Some differences in chemical composition and climate conditions result from the planets' different distances from the powerful influence of the Sun. Yet other differences seem to be due to freak accidents. Several recent discoveries suggest that immense, planet-crunching collisions were common during the turbulent youths of these worlds, and have continued to occur with decreasing frequency over their long lives. Such mammoth collisions may have given Venus its anomalous backward spin, smashed off a big piece of Earth to make our peerless Moon, and possibly created the "crustal dichotomy" that divides the northern and southern hemispheres of Mars.

To what extent are the dominant characteristics of our home planet and its neighbors due to systematic, and therefore scientifically predictable, trends related to their place of origin and evolution, and to what extent are even the most important planetary qualities due to fluke events? The answer might help us unravel our own past and assess the likelihood that planetary systems elsewhere might resemble our solar system, with Earthlike worlds.

How do planets, and planetary systems, form and evolve? We want a general answer to this question, not one that works only for our own system. We need an answer in order to determine whether our condition of existence is one of cosmic solitude or whether we live in a fertile universe with many companions for us among the stars. But we, stuck here in this solar system and barely able to leave this planet, have only three worlds with which to try to decipher everything.

Pioneer's gift of a more complete description of Venus's atmosphere greatly improved our prospects for doing meaningful comparative planetology of the "terrestrial planets," those small rocky Earthlike worlds of the inner solar system that orbit within 140 million miles of the Sun. The other major group of planets is the "Jovian planets," giant gas-balls inhabiting the outer solar system, from Jupiter's orbit nearly a billion miles from the Sun, out to Neptune's, six times more distant. When we study the physical evolution of planetary surfaces and interiors, we often include Mercury and the Moon as well as Venus, Earth, and Mars among the terrestrial planets. However, for comparative studies of Earthlike atmospheres, it is only the last three that we deal with, since Mercury and the Moon are airless. In any decent scientific study of a complex process, you would want more than three trials. However, when looking for insight into the nature and evolution of thin atmospheres around small, rocky worlds, we are restricted at present to the three planets within 50 million miles of our present location.* So let's take a global look, with comparative planetology in mind, at Venus, Earth, and Mars.

SHOULD WE TALK ABOUT THE WEATHER?

Every planetary atmosphere is a giant heat engine driven by the Sun. Weather patterns are the atmosphere's dynamic response to solar heating. The energy input is concentrated on the dayside and near the equator, where the Sun shines most directly overhead and heats most effectively. This inequity creates temperature differences that move the air around as the atmosphere "attempts" to smooth these out. How successfully it does so, and the specific patterns these motions take, depend largely on two planetary properties: the thickness of the atmosphere and the planet's rotation rate.

Venus is a good example of a planet with a thick atmosphere that very efficiently redistributes solar heat. The whole surface is at nearly the same temperature, day and night, equator to pole. An atmosphere this thick, like an ocean of water, can carry a lot of heat with it as it moves around, so the temperature differences tend to be wiped out by fluid motions. In contrast, the relatively wimpy atmosphere of Mars is much less effective at carrying heat away from the sunny places, so daily, seasonal, and latitudinal temperature differences there are extreme. Earth's atmosphere weighs in somewhere in between, as you can seen in the surface pressures given in the table on page 144. Conse-

---

*Also, Titan, a large icy moon of Saturn, has a thick atmosphere with many interesting parallels to those of the terrestrial planets.

quently, we experience more intense daily weather patterns and latitudinal temperature variations than occur on Venus, but less than those
that occur on Mars. Temperature differences generate winds, so it is
not surprising that surface wind speeds go from ferocious to moderate
to downright lazy as we head sunward from Mars to Earth and then
Venus.

In the simplest form of global circulation, solar heating causes air
to rise near the equator.* This air has to go somewhere, so it moves
north and south, away from the equator, at high altitudes. It cools off
near the poles and sinks back down, completing the cycle with a low-
altitude flow returning toward the equator. This is known as a Hadley†
circulation pattern, and each cycle of rising, falling, and returning air is
called a Hadley cell. Hadley circulation operates on all three worlds,
but forces resulting from a planet's rotation can cause more complex
motions. Venus spins so slowly on its axis that these rotational forces
are weak, and Hadley circulation dominates the redistribution of heat.
The motions in the lower atmosphere of Venus are best described as
two giant, slow-moving, planetwide Hadley cells, one to the north of
the equator and one to the south.

Compared with the slow turning of Venus on its axis, Earth
whirls like a dervish. As a result, air masses moving to the north or the
south experience strong forces that deflect them in an east-west direction. This leads to spiral, "cyclonic" wind systems. You can see these on
any given night in the weather satellite pictures on the television news.

The spinning motions so characteristic of weather on Earth are
pretty much absent on Venus. However, on Mars, with its rapid rotation and huge temperature variations, cyclonic winds are very important. Occasionally they result in giant dust storms, like terrestrial dust
devils on steroids, which can grow to engulf the whole planet. Every
few years the whole atmosphere of Mars freaks out in a global dust
storm that obscures the entire surface from terrestrial viewing until the
winds calm and the dust clears (see Figure 4.1).

Where does the superrotation of the upper Venusian atmosphere
fit into this overall comparative picture? Good question. The fact is, we
don't understand the superrotation. Our atmosphere gets its east-west

---

*Like a hot-air balloon without the balloon.

†After George Hadley, the English meteorologist who figured this out in the
1700s.

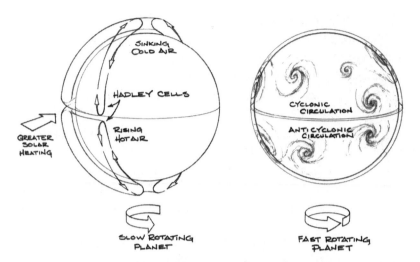

4.1 **Planetary Weather:** *The large-scale motions of planetary atmospheres are dominated by two types of circulation patterns. The drawing on the left shows the Hadley circulation that characterizes thick atmospheres on slowly rotating planets. Planets with thinner atmospheres and/or more rapid rotation are more likely to exhibit the kind of spinning, cyclonic weather patterns shown on the right. (Carter Emmart)*

winds when the rotation of Earth deflects the north-south Hadley circulation. Something like this may be happening on Venus, with the two giant Venusian Hadley cells being twisted into the superrotational motions. The problem is, Venus rotates so slowly and its upper atmosphere rotates so fast. At their peak velocities near the cloud tops, winds blow constantly toward the west at a whopping 220 miles per hour, sixty times faster than the rotation of the planet's solid surface forty miles below. What energy source could possibly be whipping the winds of Venus into this supperrotating frenzy?* It seems possible that the large amount of solar energy being absorbed around the cloud tops, by the cloud-forming photochemical reactions and by the unknown ultraviolet absorber, is contributing to this. But no one has yet come up with a convincing physical explanation or a model that can successfully duplicate this motion.

*Could it be the collective effect of all the hand-waving we do when trying to explain this?

GOLDILOCKS AND THE THREE PLANETS

Mars is essentially in the same orbit ... somewhat the same distance from the Sun, which is very important. We have seen pictures where there are canals, we believe, and water. If there is water, that means there is oxygen, that means we can breathe.
—DAN QUAYLE, former U.S. Vice President

How do the chemical compositions of these three atmospheres compare? Earth's is mostly nitrogen and oxygen, with small amounts of carbon dioxide, water vapor, methane, ozone, argon, and a host of other species. The Martian and Venusian atmospheres are both made mostly of carbon dioxide, with small amounts of nitrogen, carbon monoxide, water vapor, argon, and oxygen (Mars), or sulfur compounds (Venus). Table 4.1 summarizes the atmospheric composition of the three planets.

In this primitive, preinterstellar travel age, we should be careful about making sweeping generalizations about the evolution of terrestrial planets, since we have only three examples. All this could be a fluke. However, it doesn't take a rocket scientist to notice that Earth stands out from the CO_2-dominated norm established by our neighboring planets.

Venus and Mars have strikingly similar atmospheric compositions. This despite extremely different surface pressures—nearly one hundred times greater than Earth's at Venus, and roughly a hundred times less than Earth's at Mars. It seems that it is really Earth with the "strange brew" atmosphere. This suggests that a meaningful way to examine this trio is to look for explanations of Earth's divergence.

Table 4.1 also gives the average surface temperature for each planet. At first glance, there are no surprises: As we would expect, surface temperature decreases from planet to planet as you head away from the Sun. However, when we look at the planetary "equilibrium temperatures," the plot thickens. This is the temperature calculated from the portion of Sunlight actually absorbed and reradiated by a planet. Any light that is just reflected does not contribute to warming the planet. A darker, more absorbing object will have a higher equilibrium temperature than a lighter, more reflecting object with the same amount of solar illumination. This is why, if you are walking barefoot on a hot summer day, it is less painful to step on the white lines of a road than on the blacktop.

4.1   FAMILY VALUES: SOME CHARACTERISTICS OF EARTH, VENUS, AND MARS

	VENUS	EARTH	MARS
Major gases (%)	CO_2 (96.5), N_2 (3.5)	N_2 (78), O_2 (20), H_2O (1), Ar (.93)	CO_2 (95.3), N_2 (2.7), Ar (1.6)
Minor gases (parts per million)	H_2O (30), SO_2 (180), COS (4.4), Ar (70), CO (23), Ne (7), HCl (.4), HF (.01)	CO_2 (330), Ne (18), He (5.2), Kr (1.1), Xe (.087), CH_4 (1.5), H_2 (.5), N_2O (.3), CO (.12), NH_3 (.01), NO_2 (.001), O_3 (.4), SO_2 (.0002), H_2S (.0002)	O_2 (1,300), CO (700), H_2O (100), Ne (2.5), Kr (.3), Xe (.08), O_3 (.1)
Surface pressure (Earth atmospheres)	92	1	.006
Average surface temperature (F)	864	59	−67
Equilibrium temperature (F)	−36	−9	−80
Rotation period (relative to the Sun)	117 (Earth) days	24 hours, 4 minutes	24 hours, 36 minutes
Diameter (miles)	7,523	7,928	4,218
Mass (kilograms)	4.87×10^{24}	5.98×10^{24}	6.44×10^{23}
Average density (grams per cubic centimeter)	5.25	5.52	3.92
Surface gravity (Earth = 1)	.88	1	.38

Earth's equilibrium temperature is actually the highest of the three planets. Venus, despite its closer proximity to the Sun, has a low equilibrium temperature for the same reason that it shines so brightly in our sky: the global cloud cover reflects 75 percent of the sunlight back into space. Earth is a much poorer reflector, sending back only 33 percent of the sunlight striking it. On an airless body, like the Moon, the surface temperature *is* the equilibrium temperature. But planets with atmospheres are warmer at the surface; how much warmer depends on the thickness of the atmosphere and what it is made out of. It turns out that it is the way their atmospheres and clouds handle radiation, not simple proximity to the Sun, which largely controls the climates of these worlds. Unless you've been living in a cave (one without cable), you have heard by now of the "greenhouse effect." All three planets have some degree of greenhouse warming, due to their atmospheres, which serve as infrared blankets that prevent radiation from being emitted to space until it first pays a toll in heat (see Figure 4.2).

GREENHOUSE 101

To understand the greenhouse effect, you need to know two physical principles: (1) Everything glows, with a color (wavelength) that depends on temperature, and (2) atmospheric absorption discriminates strongly by wavelength.

Everything glows. Everything in the universe emits "thermal radiation," which results from the internal thermal motions of atoms and molecules. Since hotter objects or gases have more vigorous internal molecular motions, the energy they emit is more intense, so they emit more radiation. Of great significance for the greenhouse effect, the color of the radiation emitted from hotter surfaces is of a more energetic, shorter-wavelength variety.*

So, when it comes to thermal radiation, hotter is bluer and cooler is redder. This is why a hot light bulb radiates a yellow glow (like the Sun) but cooler stove burners are only "red hot." Objects at room temperature glow also, but since they are cooler, their thermal radiation is at longer wavelengths, out in the infrared where our eyes, narrowly tuned to sunlight, can't see. When you turn on a burner or space heater, its thermal radiation moves toward shorter and shorter wavelengths as it heats up and, as it enters the visible from the infrared, you see it begin to glow, emitting red light.

Sunlight is simply the Sun's thermal radiation. The surface of the Sun is at about 10,000 degrees Fahrenheit. Stars (or anything) at this temperature emit their light mostly in the visible part of the spectrum. (Hotter stars are blue and cooler stars are red.)

The second principle is that atmospheric gases are very particular about which wavelengths of light they let pass through them without harassment, and which wavelengths they choose to absorb. Absorption is the opposite of emission. Certain

---

*The amount of emitted radiation increases in proportion to the fourth power of the temperature. The wavelength is inversely proportional to temperature.

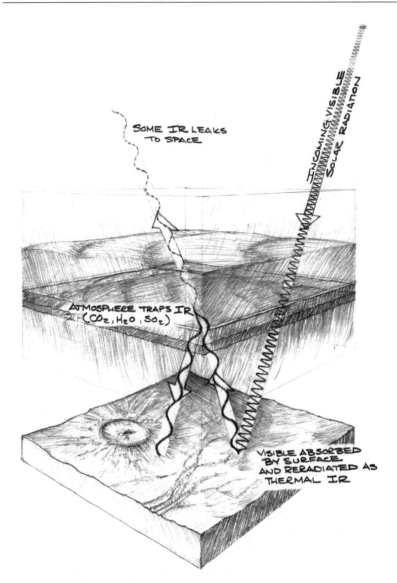

SOME IR LEAKS TO SPACE

INCOMING VISIBLE SOLAR RADIATION

ATMOSPHERE TRAPS IR (CO_2, H_2O, SO_2)

VISIBLE ABSORBED BY SURFACE AND RERADIATED AS THERMAL IR

4.2 Greenhouse Effect: *All planetary atmospheres have some degree of greenhouse warming. Visible light from the Sun travels unimpeded through the atmosphere and is absorbed by the ground. It is re-emitted as infrared thermal radiation that is absorbed by atmospheric gases, trapping the energy and heating the planet. The amount of this absorption depends on the thickness and composition of the atmosphere, and this determines how much greenhouse warming results. (Carter Emmart)* wavelengths of light—different ones for each absorbing gas—get filtered out of the radiation passing through the air, sucked back into matter, and converted back into heat, warming the absorber.

Atmospheric gases are generally transparent to visible light. They let the Sun's thermal radiation pass right on through the atmosphere, and much of it gets absorbed by the ground. This absorption heats up the surface of the planet, and so it also radiates back into space. On an airless body the balance of absorption and reradiation of thermal energy results in a surface at equilibrium temperature. But, since planets are much cooler than stars, their thermal radiation, their natural glow, is not in the visible but in the infrared. And atmospheric gases are not as kind to infrared light as they are to visible. The Sun's light makes it through to the ground just fine, but some of the infrared reradiated from the surface is trapped, absorbed by the surrounding gases. This is what leads to greenhouse warming. It won't do to have more energy coming into a planet from the Sun than is being reradiated by the planet. The planet would just keep heating up indefinitely, and this obviously isn't occurring. Instead, the surface and lower atmosphere heat up just enough so that the energy radiating back out into space, the portion not absorbed by the atmosphere, balances the solar energy received by the planet. This heating is what we call the greenhouse effect. For any amount of infrared absorption in the atmosphere, this "radiative balance" is achieved at a temperature higher than the planetary equilibrium temperature. How much higher depends on how well the planet's atmosphere absorbs infrared.

We can gauge the relative strength of the greenhouse effect on the three terrestrial planets by comparing the equilibrium and surface temperatures. Here you can see that this warming is about 900, 68, and 13 degrees Fahrenheit for Venus, Earth, and Mars, respectively. It is no coincidence that the relative sizes of these temperature differences correlate with the surface pressures, which give a rough measure of the total amount of atmosphere (although surface gravity also affects the pressure). All other things being equal, more atmosphere on a planet means more gas to absorb infrared. But all other things are not equal; the absorption of infrared radiation, and thus the climate, depends strongly on the actual mixture of gases in an atmosphere.

Earth's dominant gases, nitrogen and oxygen (N_2 and O_2), are lousy infrared absorbers. They are small diatomic (two-atom) molecules, and diatomic molecules do not absorb much infrared radiation. Even slightly more complex molecules (three atoms is plenty) make much better infrared absorbers. Because they are bigger, and "floppier," able to vibrate in many more ways, they can catch more of the infrared as it passes through, capturing its energy in the form of internal vibrations along their more numerous and flexible chemical bonds. Thus the trace gases carbon dioxide, water vapor, and methane (CO_2, H_2O, CH_4), are responsible for most of the greenhouse warming on Earth. By contrast, the CO_2 that makes up most of the atmospheres of Venus and Mars is a triatomic (three-atom) molecule that absorbs strongly in the infrared. For every ounce of atmosphere there is much more greenhouse warming on these planets.

Of the three, Earth is the only one with a climate in the appropriate range for life, or at least "life as we know it," which depends on aqueous solutions (solutions in liquid water) of organic compounds. If we did not have our "strange" (low CO_2) atmosphere, things would be uncomfortably hot for us, or we would not exist here at all, or we would be made of some other chemicals that form complex molecules at higher temperatures. The question of how it got to be this way is sometimes called the Goldilocks problem. Like the three bowls of porridge tasted by Goldilocks in the home of the Three Bears, Venus is too hot, Mars is too cold, and Earth is *just right*. When discussing the evolution of these planets, we must keep in mind the strong control of climate by atmospheric composition.

CREATING THE RIGHT ATMOSPHERE

How did Earth come to possess its unique atmospheric composition? Most studies of comparative evolution start with the assumption that Venus and Earth, neighboring planets of about the same size, began with comparable "volatile inventories," or total amounts of chemical elements capable of forming atmospheric gases. Whether this is strictly true depends on the details of terrestrial planet formation. We will return to the question a bit later when we examine modern ideas about planetary origin and early evolution. For now, if we assume that this was the case, then three obvious questions arise: "What happened to

Earth's CO_2?" "What happened to Venus's water?" and "Where did Earth's oxygen come from?"

The answers to these questions may be closely related. At present, most of Earth's carbon reserves are locked up in the solid crust in the form of carbonate rocks such as limestone and organic sediments such as oil and coal (Earth's compost heap). But you can estimate the amount of CO_2 that would be produced by converting all of Earth's carbon to this gas, and it comes fairly close to the total CO_2 content of the Venusian atmosphere. The total amount of nitrogen on the two planets is also similar. These observations have been used to justify the assumption that the two planets originally formed with roughly the same amounts of these volatile materials.* Earth's CO_2, it seems, has slowly been converted into carbonates and other crustal reservoirs. What used to be CO_2 in our air is now the White Cliffs of Dover. But why didn't this happen on Venus?

That brings us to the question of water. Evolutionary models suggest that if Venus started out with an ocean of water, it could have been lost early in the planet's history by a "runaway greenhouse." Water vapor is a powerful infrared absorber. A little water in the air can heat things up a lot. But in the presence of liquid water, if the air gets hotter, more water will evaporate. This creates the possibility of a powerful positive feedback loop: evaporating water increases the greenhouse effect, making the atmosphere so hot that more water evaporates, and so on. Any physical system like that, dominated by positive feedback, is inherently unstable. Once it gets going, there is no stopping it. Venus may have had oceans that simply boiled away, leaving large amounts of water vapor high in the atmosphere where solar ultraviolet radiation split up the molecules, allowing the hydrogen to escape into space.[†]

---

*Actually, the CO_2 on Venus is almost twice as great as the equivalent amount on Earth, and the amount of N_2 on Venus is greater by a factor of three or four. These same numbers have been used to support opposing arguments! That they are "pretty close" has been used to argue that Venus and Earth were made from basically the same material, and that they are "significantly different" has been used to argue the opposite.

[†]As I described in Chapter 3, some believe the large amount of deuterium discovered by *Pioneer* in Venus's atmosphere directly confirms this scenario.

VARIATIONS ON A THEME: FEEDBACK

Two of the most persistent themes in nature, and in technology, are positive and negative feedback. (Nature had the idea first.) Any complex system, such as a planet, a person, an automobile, or a modern building, has an array of feedback loops regulating its internal conditions. Any system with more than one element that can interact and affect each other's level of activity has the makings for feedback. Let's call these elements "thing one" and "thing two." If an increase in thing one causes an increase in thing two, and vice versa, they make a positive feedback loop. Positive feedback loops cause instability. Things one and two egg each other on to more and more outrageous behavior, and nothing stops them from getting completely out of control. A system dominated by positive feedback will tend to display extreme behavior.

If either thing responds to increased activity in the other by acting to suppress or dampen the other's activity, this creates negative feedback. Negative feedback causes stability. When thing one starts to get out of hand, thing two acts to counter this, pushing the system back toward some stable point. But this is fairly abstract. Let me give some concrete examples of positive and negative feedback:

My favorite example of positive feedback is the feedback loop that creates the sound commonly known as "feedback." Picture Jimi Hendrix jamming on his upside-down white Stratocaster guitar. Jimi's fingers (or sometimes his teeth) pull at the strings, causing them to vibrate. An electronic pickup on the guitar turns these vibrations into an electrical signal that travels down a wire into Jimi's amplifier, which does just that, amplifying the signal. The signal causes his speakers to vibrate, making waves in the surrounding air, which have the same frequency as the vibrating strings, but a higher *amplitude*. That is, the sound has the same pitch but it is louder. OK, that's how an electric guitar works, but where is the feedback? It comes in when Jimi, reaching a critical emotional peak in the song, turns his volume knob all the way up, hits the perfect note, and walks right over to his amplifier holding his guitar inches from the speaker. Now something wonderful happens. The sound waves coming out of the speaker are so intense at this point that they will cause anything near them to vibrate at the same frequencies. So the strings on Jimi's guitar start vibrating all by themselves, in direct response to the speaker, no strumming or biting necessary. But every vibration of every string is still sending electronic signals to the amp, and these signals are still being amplified, causing the speakers to vibrate with even greater intensity, which moves the strings ever harder. . . . *This* is positive feedback. The more the strings vibrate, the more the speakers vibrate. The louder the sound from the speakers, the more they move the strings. The amp goes crazy. Obviously, it can't keep getting louder forever, or it would shake the whole Earth apart. Maybe at times you thought this was happening, but it didn't, did it? The volume eventually reaches a peak because some component in the amp is maxed-out and can't get any louder. Now the tubes are really humming, rich with resonance yet on the edge of dissonance, tense yet infinitely melodic. You shouldn't necessarily try this at home. The results can be quite grating, but in the hands of someone like Jimi, who had mastered technology in the service of art, it is one of the most powerful and beautiful sounds the world has ever known.*

How about negative feedback? The common example is a thermostat. Some element senses temperature and turns a heater on, or a cooler off, if it gets below a certain point. This leads to stable behavior: the temperature hovers near the critical point where the thermostat is set and doesn't deviate greatly from it. If you made a mistake and wired your system backward so that the heater turned on instead of off when it was above the critical temperature, you would get positive feedback and it would soon get too hot. But here is a somewhat less prosaic example of negative feedback:

There is a short story by Thomas Pynchon called "Entropy" in which there is a party going on. One partygoer gets too intoxicated and heads for the shower to revive himself. He manages to turn on the water but then passes out in the shower, sitting right on the drain. The water starts to rise around him, but he is not in danger of drowning. Once the water reaches a critical level, he becomes buoyant and floats off the drain. The water runs out until the level is too low for him to float and he again blocks the drain, causing the level to rise once more. This is a negative feedback loop in which the level of water is regulated near a critical

*Shows you how beautiful technology can be, right up there with planetary exploration, heart transplants, and the Golden Gate Bridge.

point defined by the buoyancy of our wasted antihero. His behavior may be a bit unstable, but the opposing forces of his buoyancy and gravity make a negative feedback loop with a stabilizing effect on the level of the water, which should eventually revive him.

You might be wondering why this doesn't happen on Earth. After all, Earth today has water vapor and oceans, so why don't we get a runaway greenhouse? The answer is that we also have powerful negative feedbacks, which act as planetary thermostats. One of the most important involves clouds. The hotter it is, the more water evaporates. This causes clouds to form. Clouds increase the planet's reflectivity, sending solar radiation back into space and cooling us down. Negative feedback cycles like this one help to stabilize a climate system, preventing any kind of runaway. Real climate systems are usually complex combinations of many positive and negative feedbacks. That is why we are having so much trouble agreeing on how much an anthropogenic (human-made) change in CO_2 abundance on Earth will actually warm things up here.* You have to try to predict how such a change will affect all of these cycles.

If Venus did have an ocean, but lost it early, this could help to explain the dramatic difference in atmospheric CO_2. On Earth, water dissolves CO_2 from the atmosphere and turns it into carbonate rocks, a process known as "weathering." If our water disappeared, CO_2 would build up in the atmosphere. Earth avoided an early runaway greenhouse because it is farther from the Sun than Venus is and not as much of the liquid water evaporated. It seems plausible, then, that two initially similar planets at the distances from the Sun of Venus and Earth could have ended up with a dry CO_2-dominated atmosphere and an ocean, respectively. But, as suggested above, Earth's current climate depends on the small amount of CO_2 that is left in the atmosphere. Is it just a lucky coincidence that Earth's climate evolved in such a way as to maintain abundant liquid water and a comfortable environment for life? Some scientists have suggested that it is not. They have described

---

*But we *are* increasing the CO_2 content of the atmosphere, and this *will* warm Earth, so we should cut it out. Waiting for "definite" data on global warming is like jumping off a cliff to see how high it is. If we are that dumb it could easily be used as grounds to deny us membership in the Galactic Union of Intelligent Species.

schemes by which Earth adjusts its own temperature to stay warm and wet. Both nonbiological and biological feedbacks have been proposed as controlling factors in this evolution.

The story is complicated by the fact that the Sun has been getting hotter and brighter over the lifetime of the solar system. The "solar constant," the amount of radiation received from the Sun, has not been constant at all but has increased by about 30 percent, as the Sun has slowly heated, following the normal life cycle of a middle-aged star. This is easily enough of an increase to change the climates of the planets. If our atmospheric composition had remained constant, the greenhouse effect would have been too weak to keep the surface above the melting point of ice for the first 2 billion years. Our planet would have been completely frozen over for almost half its history. Yet there is abundant geological and biological evidence that it was not. For example, microfossils, traces of ancient life, have been found dating to 3.5 billion years ago, and it is generally believed that liquid water was necessary for the origin of life. Earth wasn't frozen over, but the Sun apparently was not hot enough to keep it warm. We call this the "faint young Sun paradox."

The easiest solution is to suppose that the composition of Earth's atmosphere has changed and once had a higher abundance of greenhouse gases. Given the atmospheric composition of both our neighboring planets and the similar carbon reserves of Venus and Earth, it seems quite possible that our early atmosphere was dominated by carbon dioxide. Climate calculations indicate that if the young Earth had a CO_2 atmosphere ten times as thick as the modern atmosphere, it would have warmed things up enough to offset the weakness of the young Sun. But then, as the Sun aged and warmed, Earth's surface would have heated up as well, just as your face gets hotter if you are watching a fire and it flares up. Yet the evidence suggests that, within a certain range of variations, the overall climate of Earth has remained remarkably constant. This is where the above-mentioned feedbacks come into play.

Some scientists have proposed that the CO_2 content of Earth's atmosphere is regulated over long time periods by a negative feedback loop involving the combined effects of weathering and volcanism. Here's how it might work: The amount of CO_2 in the atmosphere at any given time is the result of the balance between "sources" and "sinks." Volcanoes have probably provided a continual source of CO_2, as they do today. This supply is usually roughly balanced by CO_2 removal in chemical reactions that draw the gas out of the air and make carbonate rocks (weathering). But when the Sun was fainter, the surface would have been colder. This implies a lower rate of weathering for two reasons: chemical reactions slow down a lot when it's cold, and the winds and rain that facilitate these reactions become less vigorous when the climate cools down. In the extreme case of a frozen-over Earth, it does not rain, and weathering grinds to a halt: the CO_2 "sink" is shut down. The volcanic "source," on the other hand, is not greatly affected by climate change. Under these circumstances, volcanically supplied CO_2 would have built up in the atmosphere, causing more greenhouse warming. This in turn would heat and melt the surface, increasing the rate of weathering, sucking CO_2 back into rocks, and ultimately limiting the greenhouse.

This model suggests that as the Sun's brightness has increased, the CO_2 content of Earth's atmosphere could have gradually decreased by this negative feedback mechanism, maintaining the average surface temperature close to, but comfortably above, the freezing point of water. If this mechanism has operated over the history of Earth, it would help to explain both the relative constancy of climate over geological time and the current lack of atmospheric CO_2 in comparison with our planetary neighbors. The model also predicts that this mechanism can continue to operate for approximately 1 billion years into the future before all of the CO_2 is gone from the atmosphere and Earth is defenseless against the still-warming Sun. Nothing to lose sleep over right now.

GAIA

Or could it be life itself, the action of the biosphere, that somehow regulates the CO_2 content of the atmosphere, and controls the climate? Marine creatures incorporate atmospheric CO_2 and leave vast carbonate cemeteries of shell and coral, which eventually become limestone. Bacteria pump CO_2 from the air into the soil, accelerating the weathering process. These ideas form part of a larger set of hypotheses known collectively as the Gaia hypothesis. This seems to be defined in a slightly different way every time you hear or read it, but it goes something like this: "The combined activities of life serve to actively regulate the conditions on Earth in such a way as to maintain an optimum environment for life." Stated more simply and poetically, "The Earth is alive!" The Gaians* point out that the complex system of feedbacks determining conditions on Earth, in which biological activity plays a major role, seems to have some of the self-regulating properties of a living organism. Whether or not you consider this to mean that Earth is alive can degenerate into a dull question of semantics.

The Gaia hypothesis is hard to prove or disprove, because of difficulties defining terms such as *optimum environment for life*. It is really more a perspective than a hypothesis. But the Gaia perspective has stimulated a great deal of worthwhile scientific and philosophical study and debate over the role of life in the evolution of Earth and its atmosphere. And it has fostered an increased appreciation of the subtlety and complexity of many of these interactions. That the biosphere has had a major influence on the evolution of Earth's atmosphere is certainly true. The Gaia perspective goes further than this, however, viewing Earth's atmosphere as an integral and inseparable part of the biosphere. Life is clearly implicated in the answer to the third major evolutionary question mentioned above: the question of oxygen.

Another way to describe the uniqueness of Earth's atmosphere is to say that, compared with other planets, the mixture of gases is dramatically far from *thermodynamic equilibrium*. Thermodynamic equilibrium is the state a mixture of chemicals resides in if all chemical reactions that "want" to take place, because they lead to a lower-energy, more "relaxed" state, have taken place. If you compare the mixture of chemicals to a pendulum, thermodynamic equilibrium exists when the pendulum is hanging straight down, at rest. When any large departure from equilibrium occurs, energy is stored in the system and can be released by chemical reactions that move the system in the direction of equilibrium. Such a mixture is unstable and will change its composition drastically unless some continual input of energy keeps pushing it away from equilibrium. It resembles a pendulum suspended at an angle away from the vertical. You know it's going to swing down, toward equilibrium, unless something is continuously holding it up.

---

*A term sometimes used to describe proponents of the Gaia hypothesis.

Oxygen is an extremely reactive gas; it will combine spontaneously with nearly everything in sight. Earth's huge store of oxygen would be clearly unstable unless a dynamic process was constantly maintaining it. (Take another look at Table 4.1 on page 144.) That process is photosynthetic life. Green plants use solar energy to produce oxygen, holding up the pendulum of disequilibrium. This creates a mixture of gases like those injected into the cylinder of your car engine before it is sparked. Being far from equilibrium, this blend contains a lot of stored chemical energy that can power a car—or an organism. When we breathe, we "burn" our food with the oxygen in the air, using some of this energy to power ourselves. Without the continual conversion, by life, of energy from the Sun into chemical energy, the oxygen in the atmosphere would be used up in reactions with other chemicals, forming carbon dioxide, water, and a dilute solution of nitric acid in the oceans. This mixture of gases would be more like the exhaust from your car, having liberated its energy and come to equilibrium.

Proponents of the Gaia hypothesis maintain that a huge departure from equilibrium is a universal property of inhabited planets. This proposition, although not unreasonable, is difficult to test at present, since we don't know of any other inhabited planets. The extreme nonequilibrium state of our atmosphere clearly is due to the presence of life. Life is responsible for the large amount of O_2 in Earth's atmosphere, and also for the presence of methane (CH_4), which forms a highly unstable, nonequilibrium mixture with oxygen.* Plants, using solar energy, have converted carbon dioxide and water into food (organic molecules), liberating oxygen. Oxygen is a waste product of photosynthesis. The geologic record suggests that O_2 in Earth's atmosphere first rose to significant levels about 2 billion years ago.

Oxygen is actually a poison for organic life, because it reacts with and destroys complex organic molecules, spontaneously reversing the chemistry of photosynthesis in the quest for equilibrium. The buildup of significant amounts of oxygen in the atmosphere, largely due to the success of photosynthetic life, thus constituted Earth's first global environmental crisis. Life responded creatively by evolving the ability to

---

*If you arrive home and smell gas leaking, you should be alarmed, because a mixture of methane (natural gas) in oxygen is extremely far from equilibrium, creating the potential for an explosive release of chemical energy. This is an extreme case of the same kind of disequilibrium found in our atmosphere.

harness this disequilibrium with a controlled burn known as oxygen respiration.* Eating and breathing, we take in oxygen, turn organics back into CO_2, and derive energy from the great atmospheric battery charged up by the green photosynthesizers. (They, of course, use sunlight to turn our CO_2 exhaust right back into oxygen. Let the circle be unbroken.) This evolutionary invention unleashed enormous potential (literally and figuratively) into the biosphere. The great energy reserves thus liberated in the service of further evolution allowed the development of eukaryotes (more highly organized life, with DNA concentrated in a nucleus, rather than smeared throughout the cell), multicellular life, animals, and eventually creatures smart enough to ponder the global environment, and—let's hope—smart enough to avoid causing self-exterminating changes in it.

LOST IN THE OZONE

The accumulation of oxygen in Earth's atmosphere led to another significant development: the formation of an ozone (O_3) layer. Ultraviolet rays destroy O_2 molecules and the pieces recombine to make ozone. As you know, the ozone layer protects Earth's surface from solar ultraviolet radiation. Ultraviolet rips apart chemical bonds, destroying organic molecules and killing carbon-based creatures. The buildup of ozone shielded the surface from this biologically harmful radiation, allowing life to finally colonize Earth's land areas. Since they have almost no oxygen that is not bound up in CO_2, Venus and Mars both lack an ozone layer. On Venus, however solar ultraviolet light is effectively absorbed in the cloud tops by the unknown ultraviolet absorber and by the photochemical cycles of SO_2 that make the sulfuric acid clouds. But Mars has nothing to stop solar UV light, and its surface is constantly bathed in this deadly radiation. On Mars, without very strong sunblock (SPF 10^9), you would quickly get an all-over lethal tan.

The ozone layer causes another unique feature of Earth: the stratosphere. As you rise up from the surface of any planet with an atmosphere, the air generally gets colder. You are familiar with this if you have ever hiked or driven up a mountain on Earth, or noticed the

---

*If the disequilibrium pendulum falls all at once, you get fire or explosion, which is no way to live. But respiration allows us to ride the pendulum slowly down.

snow line or tree line on a distant mountain range. On Earth your surroundings cool by about 19 degrees per mile as you journey up from the surface until you are about eight miles high. But there the temperature starts to rise again because some of the solar ultraviolet energy absorbed by ozone gets turned into heat. The layer of the atmosphere from eight to twenty-eight miles above the surface, where temperature rises with altitude, is called the stratosphere (see Figure 4.3). Mars and Venus lack ozone or any equivalent heat source, so neither has a stratosphere.

WHY CONVECTION?

Any physical system heated strongly enough from below will undergo *convection*. This means that it spontaneously overturns, smoothing out the vertical temperature difference with rising currents of hot material and falling currents of cold material. Put some spaghetti sauce in a pan and heat it on your stove and you will see what I mean. When the lower layers start to get hot, they will become buoyant and rise, initiating convection currents. The cooler, less buoyant, upper layers are sucked downward by this, but once on the bottom they immediately begin to heat up and become more buoyant, continuing the cycle.

When you have warmer material underlying colder material, it creates an *instability*, if the temperature difference is great enough. Because hot air is more buoyant, it tends to rise. The cooler air above is less buoyant and sinks. As described above, the temperature of a planet's lower atmosphere naturally decreases with altitude. But anytime you have heating from below and cooling on top, the temperature difference becomes greater, and the situation can become unstable, initiating convection.

Think of it as a wrestling match in which the guy on the bottom keeps getting the upper hand. Being on the bottom, he is more motivated and tries just a little bit harder. The guy on top keeps getting exhausted and overconfident. He relaxes and lets his guard down. This situation is unstable. An endless string of reversals will ensue.*

---

*My twin brothers used to do this for hours on end. Even now, though they are a doctor and a lawyer, they quickly revert to this behavior if left alone together for very long.

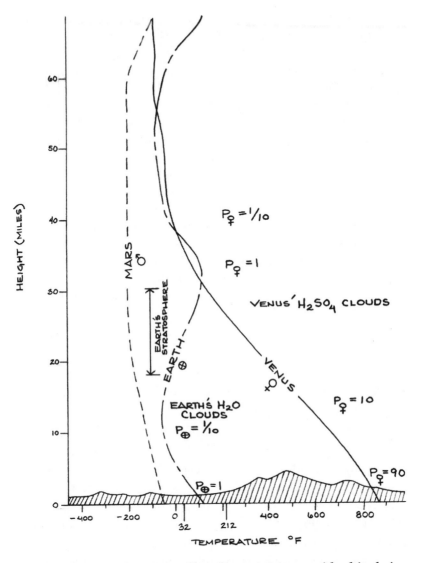

4.3 Temperature Structure: *Changing temperature with altitude is shown for Venus, Earth, and Mars. Earth's stratosphere, where temperature rises with height, is shown. Atmospheric pressures for Earth and Venus are given in units of Earth's surface pressure. (Carter Emmart)*

The lower atmosphere of every planet does the same thing. Solar energy reradiated from the ground heats the air, and it overturns (convects). Hadley circulation is atmospheric convection on the largest scale. The lower atmosphere contains convection cells of every size, from giant, planetwide Hadley cells to the small-scale disturbances that cause the turbulence enjoyed by airplane passengers. The lower atmosphere, where convection leads to constant overturning, is called the *troposphere*, because *tropo* is Greek for "turning."

But sometimes the tables are turned and warm air overlies cold air. This is called an inversion, and it is inherently stable. In this situation there is no need for a wrestling match. The warmer air likes it on top and the colder air likes being on the bottom. There is no need to move. Everyone is happy, the air is still.

Well, not everyone is happy. In urban areas, inversion layers often cause the days when air pollution is at its worst. Because this arrangement is stable, it inhibits convection, and the air stays put. Pollution from smokestacks and automobiles tends to get trapped near the ground, building up. The "brown cloud" of Denver (or L.A., Phoenix, Mexico City, etc.) forces us to stew in our own juices for a few days until there is a change in the weather and the inversion dissipates.

In Earth's stratosphere, at the level where ozone heating causes the temperature to rise with altitude, there is a constant inversion that inhibits convection and prevents vertical mixing. Such a layer becomes stratified—thus the name. The lack of convection and mixing means that particles and gases that find their way into the stratosphere stay there a long time. This is why the climate effects of powerful volcanoes that inject SO_2 up into the stratosphere last longer than those caused by weaker eruptions, which only punch up into the troposphere. The stability of the stratosphere also accounts for its dangerous accumulations of anthropogenic chemicals, such as chlorofluorocarbons (CFCs).

This sets the stage for one of the more worrisome environmental changes that humans have been causing on Earth. It is also a story about the value of comparative planetology for protecting our home world—one that ends on a tentative and cautious note of optimism. I am talking about the infamous ozone hole. The CFCs are synthetic organic chemicals containing the elements fluorine and chlorine, which were developed as "safe," "inert" compounds for our refrigerators and spray cans. But they have been eating away at our precious and fragile life-sustaining ozone layer, without which we would all get skin cancer and die, if we didn't first die of hunger from ultraviolet-damaged agriculture. You see, CFCs *are* safe and inert at the surface of Earth. What the chemists who invented these substances didn't anticipate was that before long some of their newly released creations would start diffusing into the stratosphere, where they accumulate due to the great stability of this layer. There things are different. In the presence of ultraviolet light, these chemicals are no longer inert. UV rips CFCs to shreds, splitting off chlorine and fluorine atoms, which are anything but inert. Chlorine and fluorine eat ozone. Thus the stability of the stratosphere, combined with human recklessness, contains the seeds of its own destruction.

How do we know about this? Well, we would have figured it out eventually, but the potential for harm due to these chemical reactions was first realized by atmospheric chemists studying Venus. Similar reactions occur around the cloud tops there: chlorine and fluorine compounds in the presence of ultraviolet radiation participating in chemical cycles that can, among other things, wipe out ozone. And it was Venus scientists who first said, "Wait a minute! This could also be happening on Earth," and sounded the alarm.

Our destruction of the ozone layer is increasing the level of ultraviolet radiation at Earth's surface. This and increasing acid rain are two environmental problems on Earth with strong Venus connections.* Both have been named as suspects in a recently discovered, alarming trend: this world is losing its frogs. Frog species are disappearing around the world and no one is sure why. Why is this disturbing? Aside from the fact that frogs in and of themselves are beautiful things, it may also be a bellwether, a sign of Earth's decreasing habitability. We could be next. This may be like the Old Testament plague of frogs, only in reverse, but just as ominous. Like the biblical plague, it may be a warning of more serious things to come if we don't get our house in order.

The problem is, we are not as smart as we think we are—clever, yes, but not always wise. We can invent lots of nifty things, but we can't always see the consequences of using them. It is very hard—no, impossible—to predict all the consequences of introducing a new chemical into the complex and delicately balanced atmospheric cycles of Earth. At least in this case studies of the atmosphere of Venus alerted us to the

---

*Global warming being a major third.

problem. So, if we are going to tinker with new chemicals, manufacture them in large quantities and leak them into our world (and it seems to be our nature to do so), perhaps more knowledge from comparative planetology can help prevent similar disasters.

Now here is the cautiously optimistic ending I promised: After thousands of scientific papers, hundreds of research programs, and dozens of global conferences, international agreements have been made to phase out the use of CFCs. At least for now, it looks like the ozone layer may be okay. The ozone will continue to disappear until some time in the late 90s, because the lifetime of these chemicals is long. But then, if agreements are honored and present trends continue, the situation should actually start to improve. Current projections suggest that the ozone could be back to normal in about fifty years.

The lesson is not that everything is fine now so we can start slacking off. We must continue to take our medicine if the cure is to be successful. But this does serve as a wonderful illustration that, although we have been stupid enough to create such global threats, we are capable of recognizing and fixing them collectively, globally. Who knows, maybe we can start to clean things up around here. Thank you, Venus!

THE VIEW FROM AFAR: A CASE OF VENUS ENVY?

Forgive me for the banality of this reflection, but
there is something very wrong with the human race.
—DORIS LESSING, Under My Skin

At this point in our history, when we are just beginning to detect other planetary systems around nearby stars, learn something of their nature, and fantasize about someday exploring them up close, it is fun to try to imagine what outsiders might notice about our own system on inspection.

The presence of atmospheres on the three largest terrestrial planets and the unusual nature of Earth's atmosphere would be obvious to the casual extraterrestrial observer when watching at many wavelengths over a range of timescales. The blinding reflectivity of Venus would alert them, when they were still far from the Sun, to the presence of a thick atmosphere and global clouds on that world. They would have to come in closer to detect the more subtle atmosphere of Mars, by noticing that the limb (the edge) of the red planet appears fuzzy and that over a Martian year the polar caps of frozen CO_2 grow and shrink with the seasons.

Seen in visible light, Earth's atmosphere reveals itself by its lovely hazy blue against the blackness of space. Its lively weather and life-sustaining temperature range are revealed in the rapid and complex daily movements of condensed water clouds across the globe. An infra-red comparison, revealing a CO_2 atmosphere on two out of the three planets, might be their first clue that something really strange is happening on Earth. Patient aliens observing throughout Earth's history with infrared eyes might have noticed a long-term decrease in CO_2 content. Turning their attention to the ultraviolet, they would note a dramatic increase in ozone and other oxygen compounds over the last 2 billion years. The resulting extreme departure from chemical equilibrium might alert them to the fact that Earth is inhabited or perhaps even, by their definition, alive.

Yet a puzzling set of rapid changes in the last few decades might cause them some concern about the nature of this life: they would see the clouds of Earth becoming acidified, the unique stratosphere being chemically eroded, and the anomalously low CO_2 level being slowly pumped back up toward "normal" for this solar system. The conclusion is obvious. The inhabitants of the third planet suffer from acute Venus envy. They are doing their best to transform their world into one that more closely resembles their sunward twin.

TERRA INFIRMA

Condemned to drift or else be kept from drifting.
—BOB DYLAN, "Chimes of Freedom"

The *Pioneer* mission was mostly about the atmosphere. Our ideas and models of this thick envelope have been augmented, but not fundamentally changed, since then. But *Pioneer* left us largely in the dark about the solid surface and interior of Venus—except for one experiment on the orbiter that generated a lot of heat and even shed a little light.

Pioneer's global radar altimetry map of Venus (shown in the color insert) sparked some interesting debates on the structure and motions of the interior and surface. This debate was informed by our still young but growing understanding of the way Earth works.

The story of the large-scale motions of planetary surfaces and interiors is largely one of how these bodies get rid of their heat. One axiom of comparative planetology is that smaller planets have generally cooled off and ceased most of their internal and surface activity long ago. Larger planets have been more active more recently and are likely to show ongoing signs of life.* That is because large bodies retain their heat of formation longer and hold on to their ongoing radioactive heat more easily. Physically it's easy to understand if you think of a plate of hot potatoes fresh from the oven. The small ones cool off first because of their greater surface-to-volume ratio. That is, they have more heat-radiating surface area for every pound of potato or planet. There are some exceptions to this. In our explorations of the small icy satellites of the outer solar system, we have found several nonconformist worlds that choose to ignore this rule, because of unanticipated heat sources or surface activity unrelated to internal heat supply. But the rule holds up well among the terrestrial planets of the inner solar system: Tiny Mercury and the Moon have the most ancient, long inactive surfaces. Mars is an intermediate case, being somewhat larger and having a more recently active surface, although it is pretty much dead at present except for the seasonal shifting of ice caps and windblown dust. Earth, the largest terrestrial planet, retains a hot, churning interior and a restless, shifting, bubbling, and cracking surface.

The surface expression of Earth's active interior is the constant slow bumping and grinding of rigid plates, in a process known as plate tectonics. Earth is hard on the outside where it's cool, and squishy on the inside where it's hot. This is largely due to a common property of rock and most other materials: the very strong, exponential dependence of viscosity on temperature. Viscosity is resistance to flow. Solid materials, including rock, lose their viscosity extremely rapidly as they heat up, especially when the temperature approaches the melting point. You are familiar with this behavior in butter. Freeze it and it's hard as rock;† let it warm up and it spreads easily. Rocks behave the same way. But the relatively cold outer, solid part of Earth, the litho-

---

*I mean this here in a strictly geological sense. Whether this is true in a biological sense as well is a fascinating question that we will consider in the final chapter.

†Or "slower than cold molasses." This common expression refers to the exponential temperature dependence of viscosity.

sphere,* is not made of one complete rigid shell. Instead it is broken into a dozen pieces, like shards of a giant china bowl, curved like Earth, and floating around on Earth currents. These pieces are Earth's tectonic plates, great rock slabs thousands of miles across and fifty to two hundred miles deep. They float as rigid objects: a force applied to one edge of a plate will push or pull the whole thing along. They deform along the margins, but not much in their interiors. The plates raft along on Earth's mantle, a sea of denser hot rock. The mantle is not liquid but a flowing solid, like butter or hot asphalt. These slabs play a slow motion game of bumper cars, doing most of the damage we call geology when they grind together, crumpling at the edges, or sliding over, under, or past one another (see Figure 4.4).

Plates are constantly being re-created along one edge at "spreading centers," or midocean ridges, where magma rising from the mantle cools and solidifies. The plates move slowly away from these ridges and are destroyed at the other edge, where they dive back down into the mantle—a process called subduction.

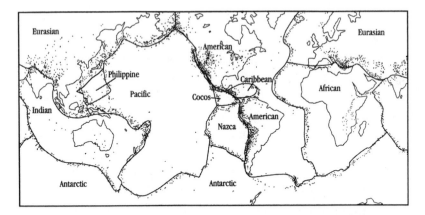

4.4 Earth's Tectonic Plates: *Earth's rigid outer shell, the lithosphere, is divided into a small number of plates that slowly move and interact, causing much of terrestrial geology. The plate boundaries are shown on this map, along with a dot for the location of each earthquake recorded over a seven-year period. Note how the earthquakes cluster along the plate boundaries.*

*So called because *litho* means "rock," and this is the sphere where rock acts like rock, not like soft butter.

The upper part of the lithosphere, made out of different kinds of rocks from the mantle, is called the *crust*. Earth's surface is divided into two very different kinds of crust: continental and oceanic. Continental crust is generally thicker than oceanic crust and, because it is made out of lower-density rocks like granite, it floats high above the surrounding ocean basins. Oceanic crust is made of denser rocks like basalt.* This *differentiated* crust is a by-product of the continuous cycling of Earth's mantle material through the plate-tectonics heat engine. Large portions of the mantle melt, rise to the surface, solidify as crust, and are later subducted, sucked down and melted again. During this, a kind of distillation occurs. Certain light elements and minerals are preferentially drawn into the melt each time, forming a low-density continental "scum" on top of the cooking cauldron of the mantle.[†] Over time this creates large areas of very different types of crust: light continents floating like barges, surrounded by lower-lying ocean basins.

Most tectonic plates have some areas of continental crust and some oceanic crust. As these gargantuan rock rafts float around Earth, following the currents of mantle convection in a way that even for Earth is still only partially understood, they undergo massive, sluggish collisions with one another. When plates collide the crust has to go somewhere. If the crust is oceanic, then one plate will be forced under another, and subduction occurs. But continental crust cannot be subducted, because it is too buoyant. So when continents collide, the path of least resistance leads to a crumpling and rising of the crust, giving rise to great ranges of highly faulted and folded mountains. One of the most spectacular current examples is the Himalaya, whose dramatic topography has been shaped by a still-ongoing collision as the Indian plate pushes up into the Eurasian plate (see Figure 4.5).

The vestiges of earlier collisions can be found scattered around Earth, mostly at the margins of continents. The Appalachian Mountains, which run along the East Coast of North America, were formed 375 million years ago when Africa and North America collided and later separated. These mountains, now worn soft and low by eons of wind and rain, were once giant and jagged as the Himalaya are today. A

---

*Continental rocks are more "silica rich." They contain low-density SiO_2. The denser oceanic rocks have less SiO_2 and more magnesium and iron.

[†]A further insult to our frail human egos from the plate tectonics revolution: We learn that the continents, where we live, are literally the scum of the Earth!

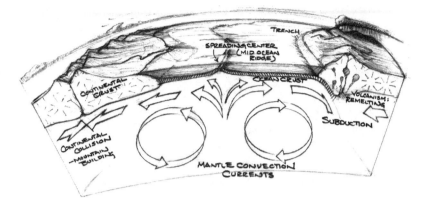

4.5 When Plates Collide: *This cartoon shows a cutaway of Earth's lithosphere and crust, indicating the major kinds of boundaries between tectonic plates and the geological structures that result. (Carter Emmart)*

few hundred million years of Earth's relentless erosion can take the edge off of any rock outcrop, no matter how massive, but this history of continental head-banging can be read in the rocks of Appalachia. They bear the internal scars of many episodes of fierce heating and crushing, just as the large-scale tectonic structure of the region tells of intense compression. If all that is not enough to convince any remaining skeptics, consider this: you can find ancient pieces of Africa among the coastal rocks of Rhode Island, Newfoundland, and other places along the East Coast.

Plate tectonics is a relatively new way of looking at Earth, but it has quickly become the unifying theory of terrestrial geology. Its power comes from its universality, its capacity to explain a wide range of superficially disparate phenomena as aspects of one vast system: the Pacific plate slides slowly northward past the North American plate and triggers earthquakes along the San Andreas fault, knocking groceries off the shelves of California supermarkets and sometimes doing far worse. The Nazca plate dives deep beneath the South American plate and partially melts, squeezing up volcanoes in the Andes. The Indian plate butts heads with the Eurasian plate, crumpling the continental crust into an accordion called the Himalaya.

By our normal standards, the motion of tectonic plates is slow. They shuffle along at an average clip of about two inches per year, about the rate at which your fingernails grow. But they make up in persistence what they lack in speed. Over hundreds of millions of years, their motions have created much of our planet's terrain.

BRUSH WITH PLATE-NESS

Many people have a story to tell about their own personal brush with plate tectonics. Here's mine: On the afternoon of October 17, 1989 at 5:04 I was in a friend's office in the Space Sciences building at NASA's Ames Research Center, where we both worked at the time. We were intent on his computer screen. (I think it was a draft of a party invitation.*) I heard a rattling sound before I felt anything. Someone seemed to be shaking, with increasing ferocity, every metal bookcase in all the adjacent rooms. I expressed some alarm but my colleague, a seasoned Californian, said "Dude, What's the big deal, its probably just another earthquake. Mellow out!" Within seconds the building started violently shaking and we were both running for the door. Yes, I know this is exactly what you are not supposed to do, but my panicked instincts favored making a run for it over hiding under a desk, and who was I to argue?

As we flew down the stairs, to the sounds of crashing furniture and breaking glass, the shaking was so intense that it was actually hard to find the next step to put your foot on. It is the only time I can remember ever feeling as though I was literally running for my life. Soon we were out front in the parking lot, catching our breath in a growing crowd of flustered people. Now the ground was doing something very interesting, sort of sloshing back and forth. Just standing on "solid ground" felt like standing on a skateboard with someone pushing it slowly back and forth. I have to admit, it was kind of pleasant, now that we were safely outdoors and could enjoy the strangeness of it. The pine trees along the parking lot were doing a line dance to the same rhythm, bending in unison first one way, then the other. And the parked cars were bouncing about like lowriders in epileptic seizure, as if impatient to join the growing traffic jam on Highway 101.

Someone had the presence of mind to turn on a car radio and see what they were saying about the earthquake. This was the most disquieting moment of the whole experience: complete

*Your tax dollars at work.

radio silence, nothing but eerie static all across the dial. At that moment, various possibilities started to suggest themselves. Was this "the big one"? Where was the epicenter and how powerful was it? Why were all the radio stations knocked out? Was San Francisco, 30 miles north, in ruins?

The silence didn't last long, and when the voices returned all they were talking about was whether the earthquake had damaged the bleachers of Candlestick Park, forcing a possible postponement of the World Series baseball game.[*] This was both annoying and comforting. Eventually reports came in that the Marina District was in flames; the Bay Bridge and a highway in Oakland had collapsed; Santa Cruz and Watsonville had been hammered. But most of San Francisco got off pretty easy this time.

This quake was not "the big one." Predictions are that San Francisco still has it coming. Still, I was lucky to be only scared to death. 62 people lost their lives, and 12,000 their homes.[†] Earth too can be hell. Events like this are calamitous for us residents of the Earth's thin skin, and CNN did its best to make this seem like a major disaster, repeating over and over again the footage of the one neighborhood in flames and the busted bridge span. This gave my parents in Boston, and many other panicked relatives the world over, the false impression that the whole city was threatened. Yet, in the long, slow movement of the Earth's great plates, this was only a tiny twitch, 7 feet of motion on a fault that just needed some momentary relief from the stress of plate tectonics.

Geology has been around for a long time,[‡] but plate tectonics, its central theory, is even younger than planetary science. Most earth scientists did not become believers in plate tectonics until the late 1960s. It is probably not a coincidence that this was the same decade in which spacecraft first started to send home images from other worlds, the decade in which *Apollo* astronauts were taking snapshots

---

[*]Friends at the game reported that the biggest challenge was trying not to spill your beer when the stands began to sway.

[†]This was called the Loma Prieta Earthquake after the location of the epicenter, 10 miles beneath the hills outside of Santa Cruz.

[‡]Perhaps one ten-millionth the age of its subject matter!

for the cover of the *Whole Earth Catalog*.* A global view was pretty much forced on anyone who was paying the least bit of attention. Resistance was futile.

But resistance there was. In hindsight it's easy to wonder about the judgment of the holdouts. Any child can look at a globe, notice the neat jigsaw fit between Africa and South America, and wonder whether there has been "continental drift." So why did it take geologists so long to accept the notion? Perhaps there is a parallel here with the stiff resistance Copernicus and Galileo met, nearly four hundred years ago, when they said that Earth moves in space. It is equally unsettling (literally) to realize that terra is infirma, that the most solid things we know, the continents themselves, are incontinent, leaking lava, forming and disappearing like clouds, slowly shifting, sliding and sinking beneath our feet. As above, so below.

Galileo's well-known (although perhaps apocryphal) remark, spoken under his breath after his forced recantation of heliocentrism, was "But still it moves!" That could also be said about the inner workings of Earth. Just as common sense tells us that the Earth is the center of the universe, it also tells us that the world is solid and not a seething, floating ball of turmoil. Is nothing sacred? Such vast stretching of the mind is always painful. Yet both of these revolutions, momentarily damaging to our false sense of place and security, expanded our consciousness immensely. Each required of us a new, enlarged perspective—in space for the Copernican revolution, in time for the plate-tectonics revolution. And each allowed us to see Earth, our home, in a much deeper way as a part of a system of worlds—an orbital system in the case of the Copernican revolution, an *evolutionary system* in the case of plate tectonics.

Part of the painful humiliation of the Copernican revolution was its great blow to the idea of Earth's uniqueness. Where does plate tectonics weigh in on this question? If we've now figured out how Earth works, can we extend the model to other planets? Or have we finally found the key to our uniqueness, a justification for our intuitive sense that this planet is special, a reason to stand up and be proud to be a terrestrial? We could test this question if there were a similar planet

---

*The exploration of the oceans, and especially the discovery of a global network of mid ocean ridges, also contributed greatly to the acceptance of plate tectonics.

nearby, about the same size, weight, and age, to examine for signs of a plate-tectonics system. Of course, there is one—the subject of this book—only it's completely socked in, obscured by clouds. Once we understood how much plate tectonics explains about our world, it became that much more important to know what goes on down on the surface of the world next door, thirty miles below the opaque, swirling acid haze.

Earth science was coming of age just as its child (with astronomy), comparative planetology, was being born. It was natural to wonder just where Earth and its plate tectonics fit in the new trans-planetary scheme of things. Picture Earth as a giant spherical frozen-over pond, not quite cold enough to make a thick solid ice sheet covering the whole pond, but with large solid sections floating around on the warm currents below. These are Earth's tectonic plates. Now picture the other "ponds" of the inner solar system, Mercury, Mars, and the Moon. These smaller worlds have less heat left below. The ponds are colder and the currents less swift. The ice is thicker and forms a solid sheet that does not move.* No plate tectonics here. Comparative planetologists refer to these worlds with continuous, thick lithospheres as "one-plate planets," to distinguish them from Earth's system of multiple, mobile plates.

What about Venus? This "pond" is basically the same size as Earth's, so, given the known relationship between size and cooling history, does this mean that its "ice" is the same thickness? Is it made of the same stuff, or does it behave differently? Is it softer, more like slush, or even harder than the "ice" here? Is the crust of this pond broken into fewer or more pieces? Is Venus a one-plate planet? And what of the currents below that drive the great rock rafts? Are they the same on Venus and Earth? If this seems like a long list of difficult questions, you've got that right. To the extent that we could make out details on the Venusian surface, we would have some basis beyond pure speculation and Earth analogy to decipher this planetary-scale puzzle.

---

*In fact, many worlds in our solar system are literally made of ice. They are the icy satellites of the outer planets. But the ice analogy is useful because frozen-over ice ponds are easier to picture than frozen rock ponds (even though we live on one).

DOES SHE OR DOESN'T SHE? PLATE TECTONICS ON VENUS

So does she or doesn't she? Does Venus exhibit plate tectonics? In a kind of planetary scale Rorschach test, scientists perused the coarse *Pioneer* radar altimetry map for signs of the familiar. This also created a useful exercise for thinking about Earth. We had to take a big step back and consider which symptoms of plate tectonics are most obvious. What would jump out at us if all we had available to study our own planet was a topography map with the same low resolution? (See color insert.) Strip Earth of its oceans, scientists argued, and the most distinctive sign of plate tectonics would be not the continents but the interconnected system of midocean ridges, also called spreading centers, where new crust is created out of magma rising from the mantle.* Was there a similar structure on Venus? Maybe, depending on whom you asked and whose papers you read.

Take a look at the map and notice the huge, Africa-sized, highland region called Aphrodite Terra that, rising two miles above the surrounding plains, lies along the equator and covers quite a bit of it. Now let your eyes travel eastward and they will hit another highlands region called Beta Regio. Keep going around the planet and a series of smaller mounds called Eistla Regio completes the circle, bringing us back to Western Aphrodite. Some researchers, connecting the dots of the *Pioneer* map, found a globally interconnected system of ridges around the equator. Was it an Earth-style system of spreading centers, midocean ridges without an ocean? Some said yes and saw this as evidence that Venus has bona fide plate tectonics. In this scheme of things, Ishtar Terra, the very large mountainous highlands region seen in the north of this map, might be the equivalent of a terrestrial continent. It would be an area of low-density rock with mountain belts pushed up by plate motion that initiated at the equatorial spreading center. But many planetary scientists were unconvinced. They looked at the same map and saw not spreading centers and continents but a less familiar arrangement of high and low areas with no obvious terrestrial analogs.

To take the argument to a higher level than "Looks like plate tectonics" "Does not!" "Does too," scientists tried some quantitative anal-

---

*We often call molten rock "magma" when it is underground and "lava" when it flows on the surface.

ysis. On Earth a cross-sectional slice of a spreading center has a unique shape that can be modeled with a simple mathematical equation based on the physics of cooling rocks. As newly made ocean floor spreads away from a midocean ridge, it slowly cools and contracts, becoming denser and thus descending in altitude. Under the assumption of uniform plate motion (the plate is moving at constant speed), this contraction produces a concave shape that can be described in a very simple mathematical formula.* The physics is straightforward and universal, and should apply to spreading ridges anywhere, on or off Earth. If the equatorial ridges of Venus really are the spreading centers of a plate-tectonics system, they must have this type of topography. Some researchers looked at cross sections across the candidate ridges on Venus, taken from the *Pioneer* map, and found that the Venusian ridges did not have the right shape. In fact, they were convex, not concave, the slope becoming steeper, not shallower, farther away from the ridge center. This meant that the Venusian highlands were not spreading and cooling like Earth's midocean ridges. The verdict: no plate tectonics on Venus.

This analysis, due to its quantitative nature, was given a lot of weight. A lot of geology is descriptive, qualitative. Planetary geologists often rely on analogy and similarities of form and structure for their explanations, and sometimes they don't use many equations. Because no two geological structures are exactly alike, geological papers are of necessity more anecdotal. This leads many scientists from the more "hard science" disciplines, such as geophysics, astronomy, and chemistry, to sometimes mistrust the conclusions of geologists. This is one of the interesting cultural clashes within the hybrid discipline of planetary science, which has been forged from several older sciences. For some, if a paper has equations in it, there is more reason to accept its conclusions. But equations, and the conclusions they lead us to, are only as good as the assumptions that go into them.

The supporters of plate tectonics on Venus found several ways around the reported mismatch between the actual and the predicted form of the ridges there. One way was to question the assumption of a uniform rate of spreading away from the ridge center. If spreading had

---

*The elevation declines proportionally to the square root of distance from the spreading ridge.

been slowing down, the prediction would fit the observed convex shape more closely. This idea nicely complemented another concept that some scientists had floated for Venusian tectonics. On Earth the creation of new crust at spreading centers is balanced by the destruction of crust when tectonic plates are subducted into the mantle. Some scientists thought the very high surface temperatures and extreme dryness of Venus would make the crust too buoyant and rigid for subduction. In that case, if a spreading center around the equator was creating crust that spread away to the north and south, the system would eventually get choked up. The crust would have nowhere to go, and the resulting back-pressure on the spreading ridge would ultimately shut it down. A topographic analysis of the equatorial highlands was consistent with this scenario of clogged-up, frustrated plate tectonics. The scenario also provided a handy explanation for the very high elevations at Ishtar Terra: presumably that northern "continent" was forced up by the pressure of the backlog of crust, which had nowhere to go but up.

Notice that the very high and low latitudes are missing on the *Pioneer* map. *Pioneer* was not able to look at the poles of Venus. This gap in the data helped keep debate alive, since the participants were free to imagine what they pleased there. Furthermore, other teams of researchers analyzed cross sections at different locations on the ridges of Aphrodite Terra and reported that they *did* match the characteristic form of terrestrial spreading ridges. Apparently, it all depended on where you looked, or what you were looking for.

HOT SPOTS

If we reject the existence of plate tectonics on Venus, that leaves us with two major unsolved problems: (1) What is creating and holding up the elevated areas (the highlands areas seen in yellow, orange, red, and white in the map in the color insert), and (2) How is Venus getting rid of her heat? One possible way to solve both problems was with *hot spots:* large blobs of hot, rising, partially molten material in the mantle that push up under the crust, releasing heat and causing elevated areas.

On Earth most of the internal heat flow is released through plate tectonics. When magma rises to the surface at a spreading center and makes new crust, it takes with it heat from the mantle. At the other end of the plate-tectonics cycle, when cold slabs are subducted, they cool the surrounding mantlelike ice cubes dropped into a drink. But Earth also has other ways to chill. About 10 percent of the total heat flow comes out from large complexes of volcanoes over hot spots. The most famous of these is the Hawaiian islands. Here, a rising plume of hot material from the mantle has been stable for millions of years, occasionally punching a new volcanic island up through the crust and the ocean. The fresh magma forming these volcanic islands carries a lot of heat to the surface, helping Earth cool its insides. As the Pacific plate slides slowly northward, it takes the islands with it, but the hot spot remains relatively stationary below. This is why the Hawaiian islands are arranged along a line with the youngest and largest ones farther to the southeast.* The crust of the Pacific plate is a conveyer belt slowly carrying the older islands off to the northwest, where they are gradually ground down by the sea. To see the Hawaiian islands from the air, or from space, is to view several million years of plate tectonics in action, and the record of a mantle hot spot that has been relatively stable over that time and that is still active today in the steaming volcanoes of the southern Big Island (see Figure 4.6).

Here is a mechanism of internal heat removal that does not depend on plate tectonics. If the crust wasn't moving, the volcanoes would just pile up in one place. So, maybe, in the absence of plate tectonics, Venus could lose its heat this way.

In fact, some places seen in the *Pioneer* topo map looked a lot like volcanic hot spots, such as Beta Regio (at 30N, 280E, in the map in the color insert). Venus seemed to be losing at least some of its heat that way. But if you calculated how many hot spots it would take to get the heat flow expected (on the basis of terrestrial analogy), Venus would have to be covered in hot spots, to an extent that could probably be ruled out even from the low-resolution *Pioneer* map. This was at best only a partial solution to the enigma of Venus's heat flow.

Most large high-elevation areas on Earth are caused by two mechanisms directly related to plate tectonics. We find the dominant

*The islands get smaller as you go north, because they are older and the sea has had more time to grind them into sand. They also change color, with the stark black of fresh, iron-rich basaltic lava flows dominating the southernmost island, and the deep rusty Martian-red colors of weathered basalts, full of iron oxides, dominating to the north.

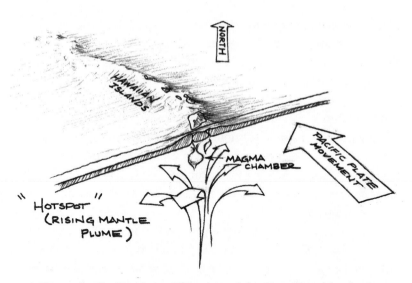

4.6 Portrait of a Hot Spot: *This view of the Hawaiian islands shows how, heading north, the islands become smaller and the topography muted as they become older and are ground down by erosion. The cartoon shows what is going on underneath: a stationary "hot spot" in the mantle is periodically punching up new volcanic islands as the conveyer belt of tectonic motion slides the Pacific plate to the north, carrying the islands away. (Carter Emmart)*

highlands on Earth on the high-floating, low-density continents. As described, the differentiation process that makes the continental crust is a by-product of the cycling of tectonic plates. The other, more localized way in which plate tectonics causes high topography areas on Earth is from the interacting of tectonic plates, the crumpling mountain building and blistering volcanics that occur along the edges, in the zones of contact between plates.

Hot spots can also create areas of elevated terrain where the momentum and buoyancy of rising hot currents in the mantle forces the crust upward. We find this type of *active support* of topography in some areas of Earth where there is a direct connection between the churning dynamics of the mantle and the surface topography. Some scientists saw evidence for active support on Venus in a *gravity map* made by *Pioneer*. This map of Venus's gravitational field was made by measuring the subtle changes in the orbiter's motion as it responded to the slight differences in gravity over different areas of the planet. To the extent that researchers could agree on its interpretation, the gravity map allowed them to actually sense density structures inside the planet and compare their distribution to the topography map. Some looked at the gravity map and thought they saw signs that hot mantle plumes were rising beneath the Venusian highlands and holding them up.

The *Pioneer* topographic mapper could not see features smaller than fifty miles across. At this resolution, geologists could not do a lot of the things they do best. For example, at higher resolutions you can look at the specific shapes of faults in the crust and determine the history and direction of the forces that caused them. Such tools would allow a more sophisticated analysis of global motions and stress patterns. With better data we could also map the distribution of impact craters, learning the age distribution of the surface. If highlands on the equator really were a spreading center, the areas around it would be made of young, freshly made crust that has had less time to accumulate impact scars than the older lands to the north and south. A map that could resolve these issues would have to wait for future missions.

THIRD STONE FROM THE SUN
One world is enough for all of us.
—STING, "One World (Not Three)"

Earth seems to be, in many respects, a most unusual planet. But, we must be careful with such statements. We are awfully proud of our vast oceans, but a Martian might feel that an oxidized red surface is where it's at, and a Venusian might think that global sulfuric acid clouds are the coolest thing. Who are we to say that our world is so special? It's good to keep this inherent lack of objectivity in mind. Nonetheless, a planetary comparison with at least attempted lack of bias does reveal a number of striking features found, at least in these parts, only on our home world.

Our atmosphere stands out with respect to both its composition and its role in the complex system of chemical, radiative, geological, and biological feedbacks known these days as "the Earth system." It is irresistible to ask which of Earth's other apparently unique features are related by cause and effect to this oddball atmosphere, and which way the causality might go (which came first). Among these distinctive features are an active hydrosphere, plate tectonics, one giant moon, a strong intrinsic magnetic field, and, of course, life.

At first glance, it seems strange that one small planet should contain so many oddities. But are these really independent developments? If some or all of them can be attributed to the same causes, this would seem less unlikely. How is Earth's unique, biologically altered atmosphere related to these other distinctive planetary qualities? Well, for one thing, our unusually large Moon may have helped maintain a climate within "healthy" limits for life. Mars has two puny moons but nothing like ours. As a result, its spin has wobbled like a top, changing the angle of sunlight at the surface and causing extreme climate oscillations over millions of years. Our Moon seems to have protected us from this hazard with the stabilizing hold of its gravity. Also, the giant impact that formed the Moon in the later stages of Earth's formation probably had important effects on Earth's subsequent evolution. As far as we know, no other planet suffered such an insult and lived to tell about it (more on this a bit later).

What about Earth's active hydrosphere, the endlessly cycling water of our world that rains, runs, and evaporates to rain again? It is clear, for reasons discussed above, that the maintenance of Earth's climate in a suitable range for liquid water is closely linked with the compositional evolution of our atmosphere. Since liquid water is absolutely crucial for "life as we know it," it is safe to say that atmospheric evolution has kept Earth alive. But how has life itself contributed to this evolution? The biosphere and atmosphere have evolved together in a most intimate fashion. But has life merely adapted to, and passively contributed to, the changing atmosphere? Or has the atmosphere been actively regulated by and for life, as the Gaians argue?

Whether the atmosphere and biosphere have had important effects on the evolution of Earth's possibly unique (at least in this solar system) style of plate tectonics is even less certain. However, two reasons that have often been proposed to explain a divergence in tectonic styles between Venus and Earth are (1) the likely lack of water in Venus's crust compared with Earth, which affects the physical properties of rocks, and (2) the large difference in surface temperature between these worlds. (We will discuss these ideas in more detail in the next chapter.) So, to the extent that life on Earth has affected, or possibly controlled, atmospheric and climate evolution, plate tectonics itself has conceivably been biologically modulated.

If you agree to that, you must agree that Earth's interior thermal evolution has been affected by its changing atmosphere and biosphere, because plate tectonics is the main way that Earth cools its interior. Even such remote quarters as the molten iron outer core, which produces Earth's singular magnetic field, may not have been immune to the modifying effects of Earth's quirky air, its unique, biologically touched, gaseous envelope.

Only one thing may be strange about Earth, with all the others following suit. But how are we to know? We cannot examine Earth's uniqueness without looking elsewhere. For we who wish to understand worlds, one world is *not* enough for all of us. Did Venus and Earth start out as identical twins, only to go their separate ways through life, or were they different from the start? The best way to know would be to ask their mother. So let's take a look at our theories of the birth of planets.

ROOTS

You are dust.
—GENESIS

We know where we're going cause we know where we're from.
—BOB MARLEY, "Exodus"

The planets of our solar system were born, 4.5 billion years ago, of the solar nebula, a flattened, spinning disk of dust and gas surrounding the young Sun. For centuries, since Immanuel Kant and Pierre-Simone de Laplace first described them in the 1700s, these planet-spawning disks have existed as theoretical, almost mythical entities. We cannot directly observe the local events of such an ancient epoch. Here we can only sift through the ashes, searching for clues. But stars are being born all over the galaxy, and we can observe similar stages of growth in some of our near neighbors. Over the last decade, thanks to the Hubble Space Telescope and impressive advances in ground-based astronomy, protoplanetary disks have become real objects, commonly observed around nearby young stars. This observational confirmation has been a major shot in the arm for theorists of planetary formation. As we long suspected, wearing a huge, flat, dusty disk is a phase that all, or at least many, young stars go through. It is a phase *we* went through. Every atom of your body, of this book, of Earth, used to be a part of the solar nebula disk. The disk didn't last long. Small bits of it rapidly coagulated, accreting into bigger bits, successively forming larger and larger objects until there were only planets and debris, and no disk was left. In less than 100 million years it was all over. We are still searching for definite signs that planets are forming within the disks we see around other stars (see Figure 4.7).

From its inception, planetary science sought general laws that could predict properties of planets from "first principles" based on such variables as size and distance from the Sun. We would like a theory that can tell us what kinds of planetary systems should develop, given some information about the "initial conditions," the circumstances of birth. One such optimistic theory, "equilibrium condensation," was first proposed in 1972. It explains how varying conditions across the disk of the solar nebula became frozen in as hereditary differences among the planets. Picture the temperature distribution in the

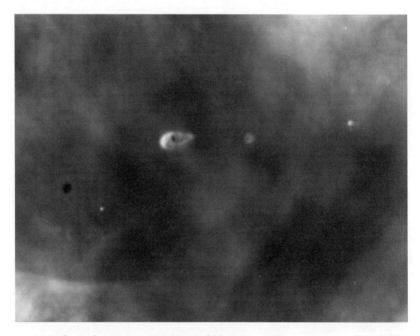

4.7 Baby Solar Systems? *This Hubble Space Telescope view of a small portion of the Orion Nebula reveals five young stars. Four of the stars are surrounded by gas and dust that are possibly protoplanetary disks. These "proplyds" may ultimately evolve into planetary systems. The Orion Nebula is 6,000 trillion miles away, and this image is 1 trillion miles across.*

solar nebula: The inner part of the disk, near the new Sun, is quite hot and everything is vaporized. At the outer fringes, far from the Sun, it is so cold even gases like nitrogen and methane can freeze. In between these two extremes is a gradient in temperature, a trend of cooling with distance from the Sun. The disk as a whole is also gradually cooling as the young Sun settles down and planets begin to form.

According to the theory of equilibrium condensation, different types of material condensed out of the solar nebula at different temperatures, and therefore at different locations. Imagine that you jump into a wormhole, travel back in time 4.5 billion years, and reenter the space-time continuum. If you entered close to the new Sun, things would be quite hot in the nebula. Around the present-day orbit of Mercury, it would be well over 1,000 degrees, and you might want to hightail it away before your shields gave out. As you gained distance on the Sun, the temperature outside your ship would steadily fall, and on your view screen you might notice that it was snowing outside. As you watched, and traveled toward the cooler regions of the nebula, you would see the snow changing colors and form.

This imaginary trip of time and space travel is just like a drive up a high mountain on Earth. As you rise past a certain altitude it starts snowing outside your car. You have passed the "snow line" where it is colder than the freezing point of water vapor. At temperatures cooler than this the thermal energy of motion in the water vapor is no longer strong enough to resist the forces of mutual attraction between water molecules. Water molecules in the air gather together and form crystals. It snows.

The equilibrium condensation theory says that a planet at a given distance from the Sun has a specific composition (is made out of certain stuff) for the same reason that the snow line occurs at a specific altitude on Earth's mountains. But the nebula contains many different kinds of vapor, not just water, each condensing at a different temperature, so there are many "snow lines." More *refractory* materials (those that melt and vaporize only at high temperatures) such as metal oxides and some silicate (rocky) materials start to snow out in the hotter inner regions of the nebula. More *volatile* (easily melted or boiled) substances such as hydrated silicates (rocks with chemically bound water) and ices begin to condense only farther out where it is colder. When these "snowflakes" of metal, rock, and ice later accreted, gathering themselves together to form growing worlds, the compositions of the resulting planets preserved this trend.

According to this theory, planets should become less metallic and icier as you go farther from the Sun, and the mineral composition of the rocks should also vary predictably among the planets. Early advocates of this simple model pointed out that it did a pretty good job of explaining density differences among all of the planets. For example, Mercury is the closest planet to the Sun and is also the densest. Equilibrium condensation explains this with the large amount of metal that would have condensed preferentially in the high-temperature zone near the Sun. The theory also correctly predicts that Mars should be significantly less dense than Earth, because it formed farther out and incorporated more of the lighter volatile materials. Since Venus formed so near to Earth, the theory predicts that their densities should be similar, which they are. This also might mean that Venus started out drier than Earth. If the proto-Venusian material was hotter, more baked by the Sun, it might have received proportionately less water in the forms of ice or water-rich minerals. This is of obvious interest for our efforts to understand the divergent atmospheric evolution of Venus and Earth. Did Venus start out substantially drier?

But, as we've learned more about planets, they've grown harder to explain. The search for simple, profound laws of planetary evolution has been thwarted by our growing knowledge of how lumpy and bumpy, full of quirks and oddities, this solar system is, and how messy and disorderly the process of planet formation may have been.

A theory like equilibrium condensation requires a fairly orderly process of planetary accumulation to maintain distinct chemical zones while planets are growing. After solid grains condensed out of the nebula and began to collide and grow into larger bodies, did these "planetesimals" stay in well-behaved, nearly circular orbits, like runners confined to their lanes at a racetrack? This would preserve the planet's chemical differences as they grew. But computer models of planetary growth raise doubts about this.

These models track the positions and motions of thousands of particles and record the outcome of their interactions. In high-velocity collisions the particles fragment, making more new particles to be tracked. Particles colliding at low velocity stick together and make larger ones. This is how planetary growth starts. The numerous near-misses alter orbits, as the participants respond to one another's gravitational pull. That's a lot of information to keep track of, so a realistic simulation requires a fast computer with a lot of memory.

As computers have become better and faster, our models of planet formation have improved rapidly, including a larger number of planetesimals and more realistically simulating their interactions. In the 1980s the new improved models started pointing consistently in one direction: Planetary growth was not orderly. It was messy and unpredictable. As planetesimals grow, their gravitational influence on other bodies also increases rapidly. When planetesimals are nearly grown into planets—when they reach around a thousand miles in diameter—things really go crazy. Their mutual gravitational interactions pull them into wild elliptical orbits and scatter them throughout the solar system. If this is a racetrack, the runners have all been dosed with psychedelics. They have forgotten the race and abandoned their lanes, and they are now gaily swinging each other around and careening about the whole field.

And so gas begot flakes, and flakes begot clumps, which begot lumps. Lumps gave birth to boulders, and boulders were small worlds unto themselves. Boulders were fruitful and joined to make bigger boulders the size of cities. And then were born the planetesimals. And near the Sun, in its great heat, planetesimals were metal. And farther out they were rock. In the farthest reaches where cold and darkness covered the nebula, planetesimals were ice. And everything was ordered and good.

And then along came mutual gravitational perturbations and screwed everything up, and chaos and confusion reigned for 10 million years.

If these model results are correct, there are three important consequences for planetary origins. First, there must have been a lot of mixing among the planetesimals that made the terrestrial planets. Simple, smooth compositional trends like those predicted by the equilibrium condensation theory are out. In the recipe for making planets, one of the last steps may have been "stir well until thoroughly mixed," so they should all be made of basically the same stuff.

Second, the details of planetary formation may be impossible to predict from first principles. Any surviving differences in composition between the planets may be due to random collisions and orbital perturbations during planet formation. The density differences among the terrestrial planets, in this view, are just the result of incomplete mixing, like lumps in batter. It may all come down to luck, the throw of the protoplanetary dice. In one famous simulation result, Mercury actually formed out near the orbit of Mars and was later gravitationally perturbed to its present location.

Third, the final stages, just before planets reached their present sizes, would be characterized by colossal collisions. Each terrestrial planet would have experienced a terrible pummeling by huge bodies thousands of kilometers across—a sort of rite of passage, a cosmic hazing before matriculating into full-fledged worlds. Such collisions would have had huge effects on the environments and physical evolution of the newborn worlds.

This last idea, a new realization of the important role that giant, catastrophic collisions may have had in determining the character of worlds, dovetailed in an interesting way with another important intellectual development of around the same time. This was a wake-up call of the wild that came 65 million years ago, but one we did not hear until the 1980s.

THE NEW CATASTROPHISM

There is no vestige of a beginning, no prospect of an end.
—JAMES HUTTON, *Theory of the Earth*, 1785

In Chapter 3, I discussed the problems of spatial and temporal variations that arise when we try to study planets using limited information. Individual planets vary from place to place and time to time, so how do we know that the places and times of our probing are not peculiar? The same problem also exists on a much larger scale: How do we know that we are observing the solar system at a "normal" time? that our system of planets is "normal" at all? or even that our part of the universe or our particular epoch—the time when we happen to have come along to briefly observe things—are at all typical of the wider and longer-lived universe?

Our notions of time have gone through some interesting transformations in the last few centuries. In our efforts to unravel the complex evolution of the atmospheres and surfaces of planets, models that invoke gradual, steady processes operating over vast stretches of geological time have been favored over theories emphasizing episodic or catastrophic events. Yet it was not always this way. In the eighteenth century the Scotsman James Hutton, "the father of geology," proposed his "principle of uniformitarianism." Hutton realized that the basic forms of landscapes on Earth could be understood as the products of familiar processes operating over vast expanses of geologic time. This was a radical new idea. Until then it was generally believed that Earth's giant landforms could not have formed without "catastrophes," huge, cataclysmic events of biblical proportions. Hutton's "uniformitarianism," advanced as an alternative to this "catastrophism," represented a major intellectual achievement. He and his countryman Charles Lyell showed that you did not need biblical floods and miracles to move mountains, but *time* was enough. Mountains could be built and then degraded by earthquakes and weather acting on previously unimaginable timescales. This put Hutton and Lyell in conflict with both religious *and* scientific authorities!

We did not fully grasp the vastness of geological time until the late-nineteenth-century discovery of radioactivity. This provided both a realization of the Sun's longevity and a means of accurately determining the ages of rocks. Several eighteenth-century natural philosophers, including Immanuel Kant, had shown that mere chemical combustion could not sustain the Sun's output for more than a few hundred years. Why, then, was the Sun shining? If it were on fire, it should have been cold and dead long ago. The realization that solar energy is nuclear allowed for a Sun that, while ultimately not a renewable resource, had burned and would continue to burn for billions of years. Along with this came an understanding of the nature of the forces binding together atomic nuclei, and governing the rates at which unstable atoms fall apart, spitting out neutrons as they go. This turned rocks into clocks and revealed the age of Earth. And guess what: IT'S OLD!!! (4,500,000,000 years)!* All human history is but a blink in the eyes of the world. The vision of Hutton and Lyell was to realize the implications of the huge difference between the scales of human and geological time. We see rivers run and flood, and we watch the occasional

---

*If any state legislature ever makes me teach "scientific creationism" in my classroom, I am also going to teach the tooth fairy, a flat earth, and the Flintstones.

movement of a boulder or bank of sand. This does not add up to the Grand Canyon on any timescale that we can easily imagine. But Earth has a lot more time on its hands.

The principle of uniformitarianism helped to achieve for our sense of time what the Copernican revolution and subsequent increases in our cosmological knowledge have done for our sense of place in the universe. Just as we no longer believe "here" to be a special location (Earth is not the center of the universe), we have learned that "now" is an incredibly insignificant stretch of time, even if "now" is taken to be the tenure of our culture or our species.* Astrophysicists and cosmologists often invoke the "cosmological principle," which states that on the largest scale the universe is homogeneous and isotropic (the same in all directions) and therefore should look the same to observers anywhere. Similarly, we have learned to be suspicious of scientific explanations that require a special vantage point in time.

Scientists strive for objectivity. We must try to prevent any hidden biases from coloring our perceptions. It is interesting to consider further how the limited scope of our existence in time might have tinted our view of the universe. The triumph of uniformitarianism was the realization that the gradual workings of familiar processes can, given sufficient time, account for large-scale geologic motions. But "familiar" processes include only those that occur so often or so constantly that they have been common even during the short time we have been around and paying attention. We don't need to invoke miracles to admit that there could be important phenomena we have not experienced in our brief glimpse of the universe. We have shown up rather late at this party, and have not been here for long. Might we have missed anything?

The uniformitarianists didn't think so. With their extreme empiricism, they assumed that any phenomena beyond Earth or unknown to their direct experience were incapable of affecting terrestrial evolution and thus could be ignored. Thomas Jefferson, on hearing a report of a meteorite fall, is said to have exclaimed, "I would much sooner believe that Yankee professors would lie than that stones would fall from the

---

*If we choose to identify with our biosphere, however, then *we* have been around for a significant fraction of the age of the universe.

heavens!" This was probably a reasonable inference based on his experience. Yankee professors do lie more often than meteorites fall.* But here direct experience was insufficient. It is now increasingly clear that strictly uniformitarian models of terrestrial evolution, by including only "familiar" processes, ignore some important features of Earth's history.

In fact, catastrophism has recently been experiencing a renaissance of sorts. The "new catastrophism" owes its existence largely to the discovery, in 1980, that Earth was whacked quite severely 65 million years ago by a large asteroid or a comet. This event ended the Cretaceous geological period and caused a mass extinction that cleared the planet of most species then living. We call this the K/T impact-extinction hypothesis.† The idea that something can come from beyond the sky and rearrange the face of the Earth, radically altering biological evolution in the blink of an eye, does not exactly fit within the uniformitarian framework of explaining everything by gradual change over eons. This discovery served as a reminder to earth scientists that Earth is a planet in the solar system and that extraterrestrial objects can have major geological implications. Much new interest in catastrophic impacts, and their role in terrestrial evolution, has ensued. We are still debating the extent to which such events have played a major role in biological evolution. The history of life on Earth may be a long series of cycles of evolution and destruction brought by impact catastrophes.

That large impacts have played a role on Earth should not have come as such a surprise. Every well-preserved ancient surface in the solar system shows a record of violent bombardment from space. Nothing is protecting Earth from this hard cosmic rain, and it is difficult to imagine that these energetic and frequent (by geological, if not by human, time) events would not have major implications for this world. One reason we were slow to realize this is that Earth is remarkably good at filling in the holes and healing the wounds left by cosmic

---

*You could substitute any profession or geographic region. The point is that meteorite falls are not among our common experiences.

†So called because this happened at—and caused, it turns out—the boundary ending the Cretaceous geological age and beginning the Tertiary. The German spelling of *Cretaceous* begins with a *K*. Using this initial distinguishes it from Cambrian or Carboniferous.

insults. There are few well-preserved craters on Earth (see Figure 4.8),* because its surface is so young and is constantly being remade by plate tectonics and worn down by erosion. Since nothing lasts long here, only the most recent craters are obvious.

After we knew that we should look, it took only a decade to find the scar left by that tremendous impact 65 million years ago. It's a buried crater one hundred miles across at the northern end of the Yucatan peninsula in Mexico, which has been named Chicxulub (after a town on its rim). Now that our consciousness has been raised about craters

4.8 Meteor Crater in Arizona: *This impact crater is one mile across and about fifty thousand years old.*

*Meteor Crater in northern Arizona is one, well worth a visit. While you are in the area, you may also want to check out Lowell Observatory in Flagstaff, where Percival Lowell mapped the "canals" of Venus.

on Earth, and we are learning how to look for them, we are finding more all the time.

At first, some scientists criticized the K/T impact-extinction hypothesis just for being a "catastrophic" idea, as though uniformitarianism were an aesthetic principle instead of an empirical statement about the ability of geologic time to make mountains out of molehills (and vice versa). But a full view of planetary history must include phenomena that are "abnormal" in the sense that they are unlikely on the timescale of our direct experience. It is certainly unscientific to reject the possibility of such events out of a philosophical or aesthetic predisposition against them.

Yet in a sense the dichotomy between gradualist and catastrophist models is a false one. Even the "uniformitarian" geological processes proceed through a series of discrete events of varying magnitude. Faults are created and mountains built in sudden, Earth-rending, quakes. Erosion does most of its damage in large storms and floods. Why do we consider a hundred-year flood to be part of a uniformitarian process, while regarding a 100-million-year asteroid impact as catastrophic? The dividing line seems to be our attention span. Fortunately, we are now rapidly extending this time base as we probe deeper into the past of our world and our universe.

The space program and solar system exploration have been essential to this form of consciousness expansion. They have given us a reliable history of the Moon, and a better assessment of the numbers and motions of the many small, crater-forming bodies in the solar system. Along with improvements in the terrestrial record (notably the convergence of paleontological and geochemical data that led to the K/T impact-extinction hypothesis), this new information has lengthened the time over which we have some knowledge of our universe, just as modern astronomy has dramatically increased the range of our spatial awareness. The principle of uniformitarianism still holds, but we must extend the definition of "normal" processes to include phenomena that, because of their very long characteristic timescales, have become known to us only by these indirect means.

The uniformitarians were correct to assume that our time is a very small piece of all time. But they erred in thinking that our time was necessarily representative. Despite the good intentions behind the cosmological principle and its temporal corollaries, it seems as much a mistake to assume that here and now are completely representative of the history and large-scale structure of the universe as it is to assume that we are at the center of the universe and that now is a special time. What are the chances that we should find ourselves in a place and time that are *exactly* average, *completely* typical, somehow representative of the whole universe? That would be a special place indeed! In a certain sense it would put Earth back in a central position, or at least a most unlikely one. I think we know better now, four hundred years after Copernicus. Universal physical laws are not incompatible with enormous diversity in the unfolding of events at particular locations. Most likely we have appeared at a place and time that are in some ways representative, in some ways unusual, and in some ways, due to the inherent spatial and temporal randomness of the universe, unique.*

Against the backdrop of this apparently reasonable assumption is the disquieting question of whether our existence itself implies a special location. Many scientists seem to feel driven to find that our emergence is a natural consequence of cosmic evolution that could have and probably has happened on any deserving planet. Yet, for all we know at present, terrestrial life may be alone in the universe.

One way to approach this haunting question is to try our best to understand the story of life's emergence and development on Earth in the context of comparative planetology. By learning *in general* how planets form and evolve in the universe and by elucidating more clearly the conditions that allowed the first self-replicating chemical systems to form here, we can begin to address the question of how common these conditions might be.

THE PERSISTENCE OF MYSTERY

But, as we have seen, the very nature of the process of planetary accumulation may defy attempts to develop general, predictive laws. In the 1980s there was a synergism between the new, sophisticated models of planetary formation—which predicted a process dominated by massive collisions in the late stages—and the new awareness of the important role of large impacts in Earth history that followed in the wake of

---

*I'm willing to bet that nowhere else in the whole cosmos is there a guitar that sounds as good as a 1965 Fender Stratocaster.

the K/T impact-extinction hypothesis. The ripples from these two new ideas were spreading, and they reinforced one another. Explaining planetary differences as being due to large random collisions became all the rage in planetary science. For a while it seemed as if every month there was a new theory explaining some mystery of the solar system by invoking a giant collision early on. How to explain the strange retrograde rotation of Venus? Giant impact. The anomalously high density of Mercury? Giant impacts must have stripped off its rocky mantle, leaving only the heavy iron core. Why is the surface of Mars divided into two very different types of landscape, with older terrain at higher elevations in the south and younger lowlands in the north? Why is Uranus tilted at such a strange angle, spinning on an axis almost perpendicular to that of all other planets? You got it: giant impacts. Finally, in one of the boldest and most spectacular applications of this principle, it was proposed that our own Moon was created by the impact of a Mars-sized planetesimal in the final stages of Earth's accretion.

I am reminded now of a conversation I once had while navigating the treacherous Los Angeles freeway system in a rent-a-car, trying to avoid impacts. When I was in grad school I attended a conference on Comparative Planetology of the Terrestrial Planets in 1985 at Cal Tech. I happened to catch a ride from LAX to the Cal Tech campus with a well-known planetary scientist, one who could be called an elder statesman of the field. He, the seasoned veteran, asked me what I intended to do for my Ph.D. dissertation. I, the nervous young grad-student trying not to embarrass myself, expressed enthusiasm for the new ideas about large impacts, and an interest in modeling their long-term effects on the evolution of atmospheres.* The advice he offered was to not get carried away and waste my dissertation on what may be a fad. I went ahead and wrote my dissertation on "Large Impacts and Atmospheric Evolution on the Terrestrial Planets." He has since made large impacts and their potential threat to humankind a major thrust of his career, publishing numerous papers and a book on the subject.

Nevertheless, the advice he gave me on that car ride was good. Really, we were both right. We need to be cautious about jumping on

---

*I have to admit that at least some of this enthusiasm probably came from the same place that makes boys and girls of a certain age love to blow things up.

bandwagons. I think of the idea of large impacts, and all their possible implications, as a long, slow, and deep thought that has been passing through the large collective mind of planetary science for nearly two decades. In this giant mind each paper published, each talk given, is a nerve cell firing. Each of these events stimulates or suppresses other neurons in response, shaping the big thought. We need the neurons of caution, inhibitors that prevent the thought from going too far.

Of the flood of ideas about large impacts produced in this collective brainstorm that swept the planetary community in the 1980s, many have stuck with us, some now seem dubious, and others still have large question marks attached to them. Perhaps the most important of these ideas, one that has steadily gained acceptance, is the formation of our Moon from debris raised when the young Earth was struck by a planetesimal as big as Mars. As fantastic as the event sounds, biblical in scale, it makes more and more sense the more you look at the evidence, and it solves a huge problem.

The Moon had always been an embarrassment, a thorn in the side of our scientific confidence, a constantly waxing and waning reminder of the incompleteness of our knowledge. We had set foot there, stolen rocks, and tasted dust, but we still didn't have a theory of its origin that wasn't easily shown to be ridiculous (much less any good understanding of those strange yearnings it induces in us).

Emboldened by the impact-friendly intellectual environment of the 1980s, a few researchers began to think what previously would have been unthinkable—that a colossal impact had created a ring of material in Earth orbit that accreted to make the Moon.* Simulations, done on the same computers and using some of the same codes as nuclear bomb design, show us that an impact can make a moon. And, it turns out, the chemical evidence—the story told in the Moon rocks brought back by *Apollo* astronauts, strongly supports this theory. The Moon resembles a cooked and distilled chunk of our planet's insides, unearthed by the biggest cataclysm our world has ever suffered. Through chemical and physical detective work we have finally and suddenly recovered memories of Earth's earliest and most severe trauma. The effects on its later development may have been enormous. If the

---

*This idea was actually proposed in the '70s but did not catch on until the "impact madness" of the '80s.

infant Earth was thoroughly baked by this event, as it seems it must have been, the terrestrial atmosphere and interior may have been greatly altered.

We don't know if Venus escaped similar traumas, but since it has no moon, we do know at least that it didn't suffer an identical one. Our theories of comparative planetology have yet to incorporate the possible consequences of these differences in early formative experiences.

By and large, planetary scientists have come to accept this wild idea as the most likely origin of the Moon. So now the Moon is a thorn in our sides in a different way. In accepting this theory, we are forced to accept the role of random, chance events in determining even some of the most basic, large-scale properties of the planets. If the Moon-forming impactor had been on a slightly different orbit, we would have no Moon and our planet might be very different, perhaps enough so that we wouldn't be here. Every time the Moon rises, it reminds us of the limits of our knowledge, and our possible knowledge. No matter how far our science progresses, the mystery will persist. (As for those strange yearnings inside, I doubt we will ever understand them well either.)

CHANCE OR NECESSITY?

You're not supposed to be here at all. It's all been a gorgeous mistake.
—SINEAD O'CONNOR, "Jump in the River"

It seems that we will never have a theory predicting in detail how a solar system arises from a disk. Just as we can predict, when we roll dice, that a seven will come up a lot more often than a three, we will learn some general principles guiding the likely evolution of planetary systems. But just as we have no power to say what the very next roll of the dice will bring, we will continue to be surprised by the result when the planetesimals roll.

Like a jazz tune where the chord changes are charted but the actual notes and their timing are left to the impulses of the moment, there may be a predictable sequence of changes on the path from disk to planets. But just as the exquisite spontaneity of jazz arises from interactions among the players, interaction among numerous orbiting and gravitating bodies seems to lead to complexity and chaos, and unpredictable behavior.

Maybe you wonder why I have waited so long to introduce the terms *complexity* and *chaos*. After all, I have been discussing the inherent lack of predictability in the process of planet-building. Isn't that what chaos is all about? Yes, sort of. But the ascendancy of large impacts in planetary science started a few years before *complexity* and *chaos* became the big buzzwords they are today. Now the chaos inherent in complex systems is used to explain irregular heartbeats, stock-market vicissitudes, and climate fluctuations. But planetary scientists did not need chaos theory to conclude that, if huge impacts were important in early planetary history, the chance location and timing of these events would have introduced a lot of randomness, a lot of *contingency* into the histories of the planets. That is, the details are dependent on when or whether specific events happened, not, as we have discussed, on nice clean and reliable evolutionary principles.

But, if the random differences were due merely to this contingency, then they were still, in theory, predictable. One could imagine a computer powerful enough to keep track of all the orbiting, gravitating, accreting bodies, and a program good enough to predict every orbit, every collision and near-miss, and the outcomes of all these events. Such a computer could predict every impact event and, therefore, numerically predict the detailed evolution of a planetary system. What chaos theory tells us, though, is that it doesn't matter how good a computer you have, that complex systems like this are inherently unpredictable. The weather on Earth is another example of a complex system, and "the butterfly effect" is often invoked to describe its inherent unpredictability: a butterfly stirring the air today in Mexico City could affect the weather next week in Boston.* Such "sensitive dependence on initial conditions" arises in physical systems that are complex, meaning that they are composed of numerous parts that all interact simultaneously, creating mutual feedbacks with unpredictable, chaotic results. These systems are also called "nonlinear," in reference to the kinds of equations needed to describe them.

---

*I recently read about a woman in Israel who sued a local television station for an inaccurate weather prediction. Most people have an intuitive sense of the chaos inherent in Earth's weather systems, but perhaps such forecasts should carry the legal disclaimer "Weather predictions are subject to sensitive dependence on initial conditions."

Chaos is inherent in the formation of a planetary system, with its many orbiting bodies all tugging on one other. And there is probably also chaos in the evolution of individual planets. A planet's interior contains heat that wants to get out, the heat left over from its formation and augmented by natural radioactivity. Like a pot left to simmer on a stove, this heat leads to the churning, overturning, and self-stirring that we call convection. Computer simulations of convective planetary interiors reveal that this process is often chaotic. The tiniest of differences in initial conditions can lead to very large differences later on. The surface geology of a planet often reflects the pattern of convection below, plate tectonics being the local example. Atmospheres are also convective systems, and their motions (the weather) are the epitome of chaos. On a longer timescale, a planet's climate depends on feedbacks among surface, atmosphere, oceans, and clouds. Feedbacks and the nonlinear relationships they create often generate chaos. These are all ways, and they are not the only ones, that chaos can influence the evolution of an individual planet and make it impossible for us to predict the details of planetary evolution from deterministic principles and initial conditions.

What all this means is that if the experiment of the solar system were allowed to run again starting from the same point, things could have gone very differently. Given random differences in conditions of origin due to the tumultuous final phases of planet formation, given the contingency on large random events early in planetary evolution, *and* given the chaos inherent in planetary formation and later evolution, we must ask: If the solar system had its life to live over again, would we find anything like our present Earth? This question should remain with us as we model, and search for, the development of Earth-like worlds around other stars.

TWIN STUDIES

Comparative planetology is frustrated by the small number of evolutionary examples we have to test our models. We are restricted at present to our own solar system, penned in by walls made of the immense distances between stars, the speed of light and the great energies required to approach it, the brevity of our lives, and the infancy of our technology. I know that patience is a virtue, but it is hard to wait patiently for knowledge that will come, if at all, generations after our deaths.

It's not fair. Could Mendel have discovered the laws of genetics if he had had only a half dozen pea plants to work with, and only an hour to observe them, rather than gardens and generations?* There is no such thing as a controlled experiment for testing anything so grand as a theory of planetary evolution. I mean, what are we supposed to do— take one very large flask; add 600 trillion trillion grams of rock, metal, and ice; heat and stir well; and then let cool for 5 billion years and observe? And for a *controlled* experiment we would have to do this at least twice, carefully changing one variable each time. For now we must content ourselves with data from a few distant laboratories, and invent theories after the fact for ancient experiments beyond our design.

The limitations of this small array of planets increases the temptation to regard Venus—with her strikingly similar size and proximity in the solar system—as a dry, lifeless control for the terrestrial experiment. And in certain respects the Venus-Earth comparison comes remarkably close to such a setup. The three most important parameters governing a planet's evolution are probably size, composition, and distance from its star. Venus, being the most similar to Earth in all of these properties, is a natural laboratory to test our knowledge of Earth and planetary evolution.

Even given these similar initial conditions, should we really expect Venus and Earth to be all that similar, if historical contingency plays a large role in determining planetary nature? Planets, and perhaps entire planetary systems, may resemble people more than atoms and molecules. They are individuals. Each one is different, and certain aspects of their personalities result from chance experiences. Earlier I described the unusual, sometimes troubled, but often fruitful cultural mix of planetary science. Chemists and physicists are used to constructing deep and reliable theories that explain the relatively simple, predictable, and uniform objects of their study: atoms, molecules, and other well-

---

*In this case Mendel probably would not have discovered the gene, but eventually, with the techniques of modern molecular biology, we would have figured it out. In the absence of a statistically large empirical database, we need to probe deeper, seek the mechanism, to find the way things work. The analogy in the field of planetary evolution is perhaps the isotopic analysis of rocks and meteorites, which tells us so much about the history of these objects and the solar system that spawned them.

described physical systems. Geologists are more comfortable with complex and intractable things like landscapes, mountains, and rivers. Deep physical principles describe the general properties of rivers and mountains, but none will account for each bend and crag. For planets, both perspectives are valuable. We can still search for theories that explain their large-scale features while recognizing that they are also individuals whose personalities are the sum of their experiences. The disciplines joined in the shotgun wedding of planetary science provide not only the right combination of expertise but a healthy balance of perspective between these different ways of thinking about scientific problems.

For example, although recent theory suggests that equilibrium condensation in its pure form was too optimistic, some of its predictions do seem to have survived the tumult of planetary accretion, and the theory, in modified form, survived the tumultuous 1980s. Although the details of terrestrial planet compositions may not follow a simple, predictable trend with distance from the Sun, equilibrium condensation does predict small rocky terrestrial planets orbiting near the Sun, and giant gas-ball jovian planets farther out. This theory does a very good job of explaining some of the large-scale properties of our solar system, even as it leaves the details of individual planets to chance. This is a good candidate for a planetary formation theory that could be universal: it is quite possible that someday, when we do have a number of planetary systems to examine, many of them will have this same overall structure. This may be a common story. But the details will always be different and surprising.

Many of the questions raised above boil down to this: Did our system of worlds form and evolve the way it did out of necessity or by chance? This reminds me of the long debate about "human nature." Are we who we are because of programs written into our genes, or are we mostly a product of our experiences and environments? To both questions we can reply that it is surely some combination of the two. But what is the balance?

One of the most important tools for exploring the nature-vs.-nurture question in humans is the use of "twin studies," especially studies of identical twins who have lived their lives in different environments. Because they are genetically identical, any differences between them in appearance or personality most likely result from their separate experiences.

The terrestrial planets are very much like a group of siblings, a litter of four, born together from the same dusty disk. Mercury is the runt of the litter orbiting close to Mommy's skirts. Venus, Earth, and Mars are the closest siblings. They have remained near each other since birth. The giant, Jovian planets and distant, icy Pluto are more like cousins. They share some genetic material with the terrestrial planets. If you go back far enough, we all come from the same molecular cloud. But their parent bodies were fundamentally different.

Are Venus and Earth twins separated at birth? How have their different environments and experiences affected their development? Remember that not all twins are identical. They may share some genetic material, but not all. And even identical twins may be born with somewhat different weights. Just as human twin studies help sort out the influences of nature and nurture, perhaps this planetary twin study can help us resolve between the roles of chance and necessity in determining the fates of worlds.

Here are two planets that "should" be similar, given their similar size and apparently similar composition. It is true that Venus is closer to the Sun, but that's okay. In controlled studies you are supposed to change one variable and leave everything else the same. So the question here might be, can the divergence of Venus and Earth be explained by their 30 percent difference in distance from the Sun, or their slight size difference? Or must we invoke contingency and chaos? Venus will not tell us the answer to these questions, but she is helping us learn how to think about what it takes to make Earthlike planets, and thus we will be a little less naive as we begin to explore the galaxy.

Pioneer Venus provided us with a detailed portrait of the atmosphere of Venus, and this solidified our realization that Venus, the last best hope for a nearby Earthlike world, was very different from Earth in some major ways. But what of the surface of Venus? Would it reveal the familiar, telltale signs of plate tectonics? After *Pioneer* we had some general impressions, an out-of-focus picture, a crude map sufficient to fuel a debate about Venusian plate tectonics but not to resolve it. We still didn't know what the surface was like, but we did know how to find out. The answer was radar imaging. *Pioneer*'s crude topo map gave just a hint of what we could do using radar. We knew we could take actual pictures of the surface from a spacecraft with a radar camera.

Once again the major obstacles were politics and bureaucracy. Plans for such missions, in the works since the early 1970s, were thwarted for many years by cancellations, delays, and cutbacks. Same old story. But members of the Venus science community were persistent. They kept improving, streamlining, and trimming down the concept until they gave Congress an offer it could not refuse, a spacecraft called *Magellan*. *Magellan* was, quite literally, a relic of a waning age of more ambitious and confident space exploration. To save money it was constructed of leftovers and spare parts from spacecraft built in that more extravagant era. The earliest designs were, like *Pioneer*, comprehensive and ambitious, bristling with instruments. These missions never flew. The instruments were discarded one by one as the planetary scientists, like desperate ballooners shedding ballast, tried to keep their mission aloft. In the end, only the radar imager remained. As the 1980s ended, *Magellan* was in interplanetary space, on its way to a five-year mission to explore a strange new world. It carried with it our high hopes that the clouds that had taunted us for centuries would at last be parted, and a promise that, if it worked as hoped, 1990 would mark not just the start of a new decade but a whole new era in our understanding of Venus.

5

long-lost sister: magellan and the rediscovery of venus

This is the story of how we begin to remember.
—PAUL SIMON, "Under African Skies"

A new worldview is being born right now.

Such claims are often made and are usually specious, but in this case I mean not our world but Venus, the world next door. *Magellan* has so radically and suddenly removed the shroud that it would be surprising if a revolutionary change in our Venus story did not follow. This will ultimately lead us to a revision and clarification of our Earth story, a revolution in planetary self-knowledge. We have watched Venus metamorphose from a goddess deeply involved in human affairs to our world's heavenly twin. We've relived the disappointment of finding that "Venus is hell" and found a new appreciation of how studying its atmosphere can help us recognize, understand, and rectify some of our dangerous tinkering with our own atmosphere.

Humans have now sent more robot spacecraft to Venus than to any other planet (not counting our Moon). The most recent of these has done more than all others combined to transform Venus, in our minds, into a *place*. The latest chapter, for now, in the saga of Venus exploration has finally given us a clear view of the solid surface. We can map and explore the mountains and valleys of our goddess, our hell, our experimental "control." And what we find is that in some very important ways Venus *is* Earth's twin after all.

RADIO-FREE VENUS

It's not quite true that we had no images of the Venusian surface before we sent spacecraft there to work their magic with surface cameras and orbital radar. There was one oddly shaped porthole through which we peered for decades, gaining a somewhat distorted and twisted view, always of the same area of the surface.

For so long the world next door lay hidden from our view, fenced off by thick atmosphere and clouds. Beyond this spectral fence lay the kingdom of Venus, if only we could find a way to see it. We slid along the spectrum seeking a place where we could peer through to the ground. In the ultraviolet we saw interesting cloud markings but nothing anywhere near the surface. In visible light we saw nothing but bright reflecting clouds. Through the whole wide infrared region we searched in vain; the thick CO_2 atmosphere obscured our view.* Then, finally, out where the waves are measured in inches, our search paid off. Gaps began to appear. The atmosphere and clouds are almost completely transparent to electromagnetic radiation with wavelengths between one and ten inches. Here, in the region of microwave and radio, we could finally see clear to the surface.

In Chapter 2 I discussed how the declassification of radar after World War II gave birth to the science of radio astronomy. Since radio waves can travel unimpeded through the Venusian atmosphere and clouds, this was a major boon for Venus science. In 1956 it allowed us to discover the anomalous microwave radiation that ultimately precipitated

---

*As discussed in Chapter 3, we found some intriguing cracks years later in the near-infrared part of the fence, a few loose pickets through which we could sense the lower atmosphere and even vague forms on the surface.

Venus's fall from grace, from twin to hell. And a few years later radio observers were able to resolve the long debate over the rotation rate of the solid surface. Recall that earlier telescopic attempts had produced two competing popular ideas for the length of day: either it was closely linked to Venus's orbital period of 225 days or it was nearly identical to Earth's twenty-four-hour day. Since the rotation rate plays an important and somewhat surprising role in the following story of early radar discoveries, let me refresh your memory with the following excerpt from C. S. Lewis's 1944 morality play *Perelandra,* which takes place on Venus. The protagonist is about to depart on the first human voyage to Perelandra (Venus), and he is discussing with his friends what he might encounter:

> "Our astronomers don't know anything about the surface of Perelandra at all. The outer layer of her atmosphere is too thick. The main problem, apparently, is whether she revolves around her own axis or not, and at what speed. There are two schools of thought. There's a man called Schiaparelli who thinks she revolves once on herself in the same time it takes her to go once around [the Sun]. The other people think she revolves on her own axis once in every twenty-three hours. That's one of the things I shall find out."*
>
> "If Schiaparelli is right there'd be perpetual day on one side of her and perpetual night on the other?"
>
> He nodded, musing. "It'd be a funny frontier," he said presently. "Just think of it. You'd come to a county of eternal twilight, getting colder and darker every mile you went. And then presently you wouldn't be able to go further because there'd be no more air. I wonder can you stand in the day, just on the right side of the frontier, and look *into* the night?"

Finally radar told us how fast Venus spins. In 1961 scientists using the 210-foot Goldstone radio telescope in the Mojave Desert of California bounced a radar beam off Venus, and analysis of the return echo revealed that it rotates quite slowly. Later refinements of this technique

---

*On Venus our hero finds a "normal" day length, and a planetary paradise mostly covered with oceans of water and populated with all manner of wonderful creatures. And who does he encounter there but a great physicist who has been possessed by Lucifer himself!

showed that the planet turns on its own axis in 243 days.* Unlike Earth, Mars, and all other planets but Uranus, Venus rotates retrograde, that is, in the opposite sense of its orbital motion, turning clockwise as seen from above its north pole. The period of this slow backward rotation is slightly longer than the Venusian year (orbital period around the Sun) of 225 days. The combined effect of the rotation and orbital period makes for a solar day (from one noon to the next) of 117 days. In the historical discussion of rotation rates in Chapters 2 and 3, I gave you this answer, but I saved the punch line for now, because of its impact on the history of our radar explorations of Venus. There is something quite odd about this rotation rate: once again, for whatever reason, Venus has chosen a time interval curiously linked to Earth.

ANOTHER STRANGE PENTAMETER

Recall that although Venus orbits the Sun in 225 days, she passes the Earth only once every 584 days, since we are moving on a slower orbit in the same direction.† We have already discussed the strange five against eight resonance, well known to the classic Maya, between this orbital period and our terrestrial year, producing the five sky paths that repeat every eight years. Now, here is another strange fact of planetary numerology: within this time period of 584 days neatly fits 5 of these (117-day) Venusian days. More precisely, there are 5.001 solar days on Venus between one inferior conjunction and the next. This spin resonance is completely separate from the orbit resonance; they are two independent coincidences. Together, they ensure that Venus spins almost exactly twenty-five times over the eight-year cycle of five repeating paths so dear to the Maya. This adds an additional rhythmic counterpoint, an *agogo* bell ticking off a measure of five for each drumbeat of Venus's slower orbital five, which itself repeats in time with Earth's eight. It would have delighted the Mayan astronomer-priests.

There is irony here, given our long record of false consensus on rotation rates. This history reveals obvious desires to find a rate linked

---

*In this discussion, "days" means Earth days unless I specify otherwise.

†This interval, the period with respect to the Earth, is called the synodic period.

somehow to Earth, or to Venus's orbital period. The funny thing is, when we finally did establish the rotation rate, using radar, we found that it is linked both to Venus's orbital period *and* to Earth, since it is linked to the orbital period *with respect to Earth*. But there is absolutely no way that early telescopic observers, seeing only the wispy and often imaginary markings in the clouds, could have caught any hint of this.

Like a machine with gears almost perfectly meshed to Earth, Venus spins around its axis five times between every close encounter with us. But why? For some reason that we haven't been able to determine yet (but there should be one), Venus's spin, like its orbit, also sustains a rhythmic relationship to Earth's orbit. Really. I'm not making this up. It's published in reputable journals. These obvious synchronizations with no discernible explanation are a bit bothersome.

Could it be just a coincidence? Of course it could. It's even off by a tiny bit.* There are some weird coincidences in nature that we just have to live with, even in the basic structure of the solar system. Consider the fact that the Sun and Moon are the same size as viewed from Earth. The Sun is four hundred times larger in diameter than the Moon, but also four hundred times farther away, so the Moon covers it like a custom-designed sunshade during a solar eclipse.† This is one coincidence we should be thankful for, since total solar eclipses are among Earth's greatest sights.

But scientists don't like coincidences, and the more unlikely they seem, the more we scramble for a logical, causal explanation. So we have some theories about how the spin of Venus may have been pulled into sync with Earth. Unfortunately, they don't really work, at least not yet. The most obvious possibility is that Earth's tidal forces could have influenced Venus's spin. Earth and Venus probably raise small tidal bulges on each other, just as Earth and the Moon do. These are just like

---

*The difference between the synodic period of Venus and the period required for an exact resonance with Earth is about 0.17 days out of 584.

†This is true right now, anyway. The Moon is slowly receding from Earth, and billions of years ago it appeared larger in our sky. The *Apollo* astronauts left laser reflectors on the Moon, so we can measure its distance very accurately. These measurements confirm that it is indeed fleeing at a rate of about two inches per year, and its apparent size is very slowly shrinking. The "coincidence" is partly one of timing—when we happened to come along and notice.

Not in Kansas Anymore: *These six photographs were taken on the surface of Venus by Soviet landers. From top to bottom:* Venera 9 *(1975),* Venera 10 *(1975),* Venera 13, Venera 13 *(1982),* Venera 14, Venera 14 *(1982).*

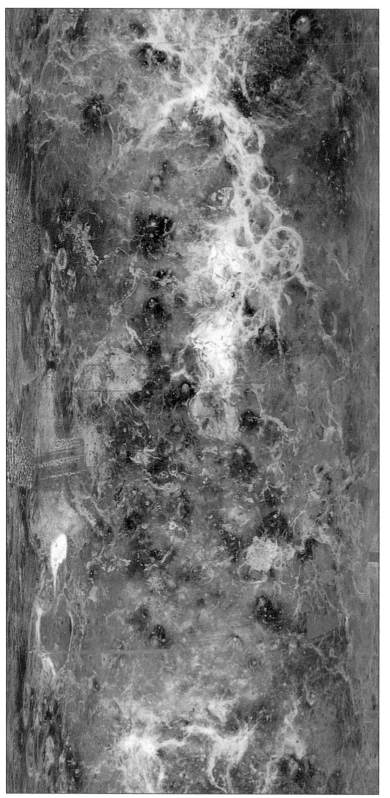

The Big Picture: This is a mosaic of the entire image set obtained by Magellan at Venus. In this cylindrical projection, the right and left edges are at 240 degrees east longitude. The bright area in the upper left is Maxwell Montes. Extending along the equator to right of center is Aphrodite Terra.

The Giant Volcanoes of Eastern Aphrodite: This sixteen hundred-mile-wide image shows the shield volcanoes Maat Mons (lower left) and Ozza Mons (center). Maat Mons, towering twenty-five thousand feet above the plains, is the largest volcano on Venus. Fresh looking lava flows can be seen extending hundreds of miles down the flanks and onto the surrounding plains. The bright lines cutting through Ozza and the southern flanks of Maat are faults caused by rifting of the crust.

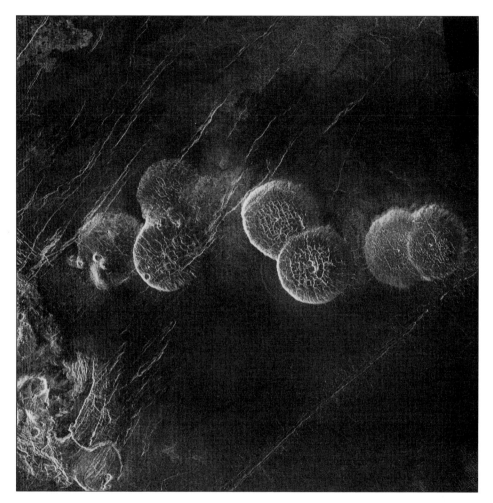

Pancake Domes: *These seven steep-sided domes, located along the eastern edge of Alpha Regio, are each about fifteen miles across and over two thousand feet high. They seem to have been formed by eruptions of very viscous lava which piled up instead of flowing long distances.*

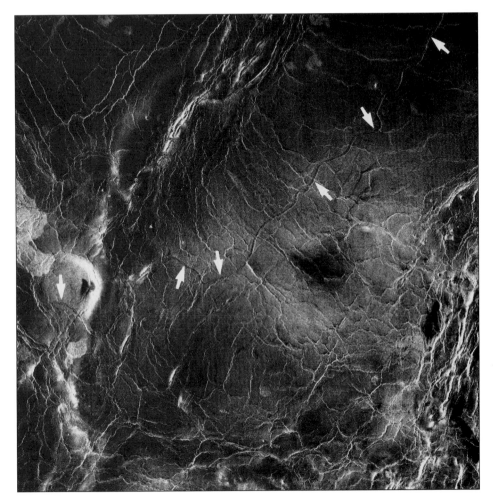

The Longest River: *Baltis Vallis, the longest known river channel in the solar system, meanders over the plains of southern Atalanta Planitia. This image shows a 360-mile segment of a channel that is more than forty-two hundred miles long. The average width is about one mile.*

Strange Volcano: *This volcano, located southeast of Artemis Chasma, is unlike any other found on Venus (or any planet). The lava flows are very rough and quite thick (about two thousand feet). They radiate from a central dome sixty miles across, and extend to a distance of up to two hundred fifty miles.*

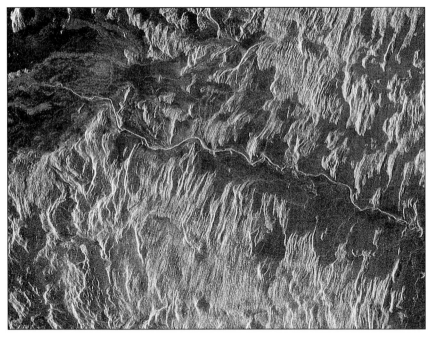

A River Runs Through It: *This sinuous channel near the western end of Ishtar Terra is clearly of volcanic origin, but it flows like a terrestrial river for more than one hundred miles. The origin can be seen in a circular depression at the lower right of the picture.*

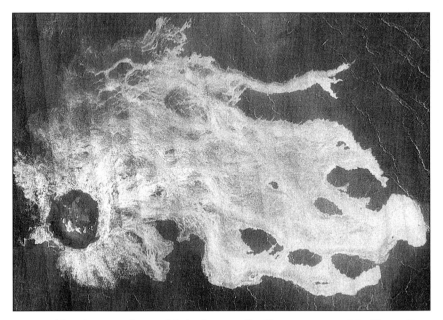

Crater With Giant Outflow: *This is the 43-mile diameter crater Markham. Large craters on Venus are often found with voluminous bright outflows extending from them. These extended crater-generated flows are not found on any other planet. Their origin is mysterious, but it may be that the high temperature facilitates melting of surface rocks from the energy of impact. The outflow in this picture extends two hundred thirty miles from the crater.*

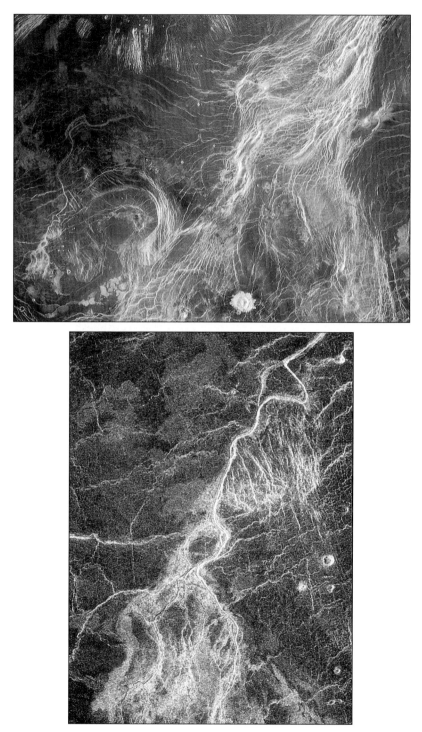

Delta of Venus: *A river delta can be seen at the end of this channel that winds through plains south of Ishtar Terra. To the east is a large belt of ridges. This image is three hundred eighty miles across. At bottom is a detail sixty miles across, focusing on the delta structure. Places where the stream has wandered and flooded its banks can also be seen.*

Volcanic Menagerie: *This portion of Atla Regio contains many types of volcanic features and superimposed tectonic features. The image is two hundred seventeen miles across. Lava flowing from circular pits or linear fissures has formed flower-shaped patterns in several areas. Numerous fractures and linear valleys cut across the volcanic deposits, indicating that the tectonism post-dates most of the volcanism here.*

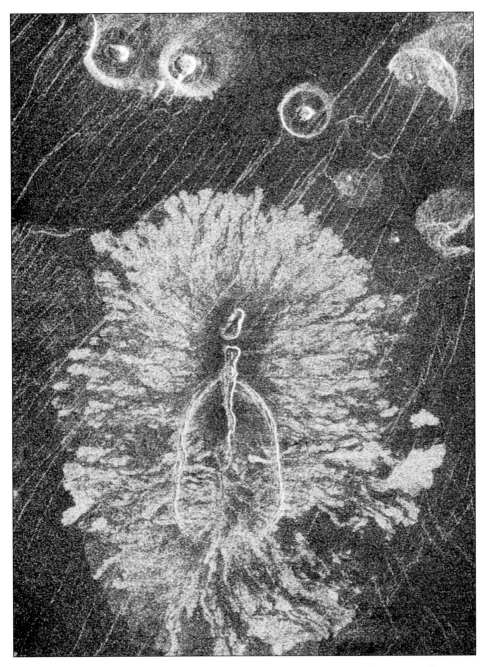

Anemone in a Sea of Basalt: *This detail from the previous image shows a petal-type or "anemone" volcano. Bright lava flows radiate from an elongated fissure. The feature is 25 x 35 miles.*

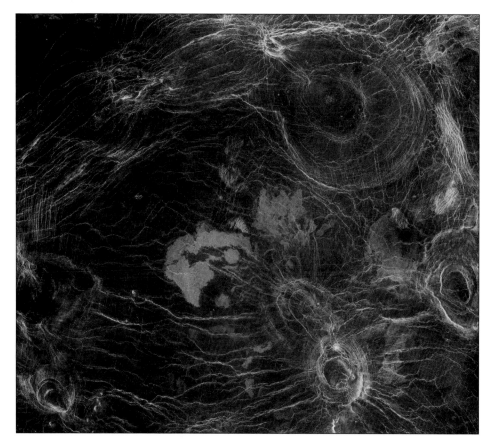

Arachnoids: *These web-like features are found only on Venus. They are made up of radiating patterns of cracks, probably caused by upwellings of magma beneath the crust. The arachnoids seen here range from thirty miles to one hundred forty miles in diameter.*

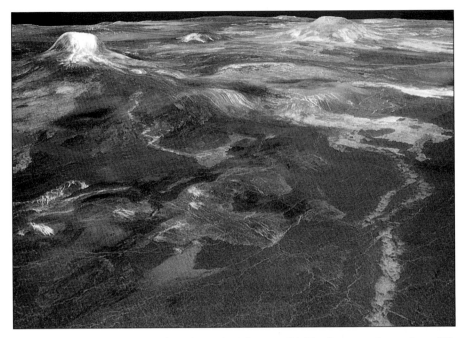

3-D View of the Volcanoes of Eistla Regio: *The two shield volcanoes pictured are Sif Mons (left) and Gula Mons (right). The distance between the two is four hundred fifty miles. Lava flows extend for hundreds of miles across the fractured plains in the foreground. This 3-dimensional perspective view was made by combining* Magellan *imaging and altimetry data. The vertical scale is exaggerated ten times to bring out the details of the topography.*

3-D View of Eve Corona: *This vertically exaggerated perspective image shows a 200-mile diameter corona on the southwest ridge of Alpha Regio. The roughly circular form and concentric ring of ridges are typical features of coronae. (Daniel Janes)*

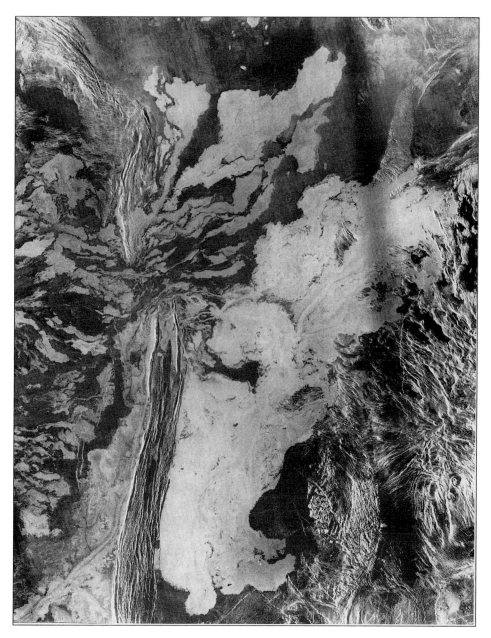

After the Flood: *This massive lava flow is composed of radar-bright and radar-dark lavas that have flowed over four hundred miles east from their source area. The flows encountered and breached a large north-trending ridge belt and ponded to the east, forming a vast, bright deposit. The total area of this flow field is roughly 200,000 square miles.*

Great Rift: *This is Devana Chasma, the giant rift valley in the center of Beta Regio which has been compared to Earth's East African Rift. The image is three hundred twenty miles across.*

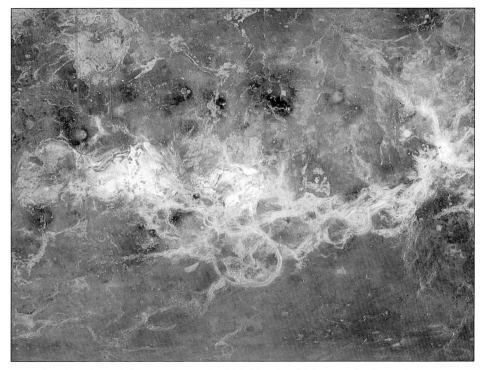

Aphrodite Terra: *This continent-sized highland, which extends along the equator for 10,000 miles, has long been at the center of debates about global tectonics on Venus.*

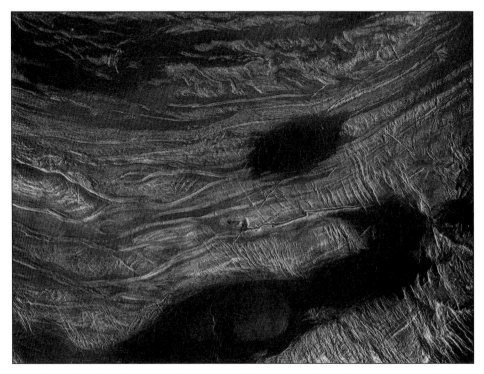

Ridges of Ovda: *This image shows part of the northern boundary of Ovda Regio, in western Aphrodite Terra. The scene consists largely of low, linear ridges. These ridges, five to nine miles wide and twenty to forty miles long, lie along a 100-mile wide slope where the elevation drops two miles from Ovda Regio to the surrounding plains.*

ocean tides of water but smaller because they are raised in more resistant rocky material. Earth could pull Venus into line by tugging at these bulges. In fact, this tidal tugging does explain why our Moon has a synchronous rotation (turning so that one side always faces us). But the strength of tidal forces falls off very rapidly with distance. Calculations have shown that Earth's tidal forces on Venus are much too weak, by many orders of magnitude, to have the required effect.

I don't understand this strange synchronicity, and I don't think anyone does. But I don't think it's a coincidence. There probably is a causal explanation that we have not yet discovered, possibly linked to events during the time of formation of these worlds.

Many people have marveled at the fact that a certain style of guitar playing from the West African country of Mali sounds so much like American Delta Blues.* Is this just a coincidence? Is it an example of convergent cultural evolution—the same innovation appearing independently in two places? Is it a manifestation of some "hard-wired" musical intelligence? Perhaps, but more likely these styles are linked via the slave trade and mutated cultural memories that resurfaced across the Atlantic. Ancient links can manifest themselves as similarities that seem like coincidence. Such may be the case with orbits and rotations of Venus and Earth.

Whatever its cause, the Venusian rotational lock with Earth means that the same hemisphere always faces us when she draws nearest.† This has had major consequences for our exploration of Venus.

The next step in our radio study was to make images of the surface. Radar imaging works essentially like flash photography. You send a blast of radio waves at your subject and gather the return pulse into an image. The quality of radar imaging is highly sensitive to the distance between subject and observer. We can get decent radar views of the surface of Venus only right around inferior conjunction. As she swings briefly between us and the Sun, she comes close enough for us

---

*The American musician Ry Cooder has said that the guitar style of Malian master Ali Farka Toure "sounds like the Blues played backwards"!

†Consequently, if you were on Venus and could see beyond the clouds, Earth's motions would be oddly linked to your geographical location. Certain longitudes would be no good for Earth watching, and at closest approach Earth would always appear directly overhead at the same spot on the planet's surface!

to snap a few pictures before receding again. This fact conspires with the strange rotational lock to ensure that radio telescopes stuck on Earth always see the same side of Venus.* And we can get decent views of only part of this hemisphere, because radar image quality depends on the angle between the radio beam and the surface you are imaging. The angle is nearly perpendicular at locations near the Venusian equator but much shallower at higher latitudes. As a result, ground-based radar images, especially the early ones, looked like photographs taken with a distorted lens. Nonetheless, beggars can't be choosers, and we were grateful for any surface images at all.

By 1970, we had obtained images of Venus from several radiotelescopes. The largest and most powerful of these is a one-thousand-foot metal bowl at Arecibo, Puerto Rico. In early images from Arecibo you could make out features larger than thirty miles across (see Figure 5.1).

Radar images differ in some important respects from our familiar photos taken in visible reflected light. The roughness of the surface determines how much of the signal bounces back to the receiving antenna; rough areas look bright and smooth areas look dark.† Brightness also depends on the inherent radar reflectivity of the surface, which in turn depends on its small-scale texture and the type of rock there. Interpretation involves trying to separate out these factors.

Scientists assumed that the most prominent bright areas seen in these first images were rugged, perhaps elevated, terrain, and gave them the nondescript names Alpha and Beta. (This was before the convention of female place names on Venus was established.) Over the intervening decades, Alpha and Beta have gradually been transformed from vague, bright spots on a ghostlike image to real places with characters and histories. The new maps of Venus had the zero meridian of longitude running through Alpha, a distinction belonging on Earth to England's Greenwich Observatory. At Alpha's location on Venus, Earth is always directly overhead about a month after inferior conjunction.

At every conjunction new radar images were made using improvements in technology and technique that had become available

---

*Since the rotational sync is not exact, Venus actually does turn, very slowly, for Earthbound radar observers. But don't hold your breath: from Earth's point of view, one rotation takes about twenty-three hundred years.

†An analogy to this can be seen in sunlight reflecting off water. When wind makes the water rougher, you see more bright reflected light.

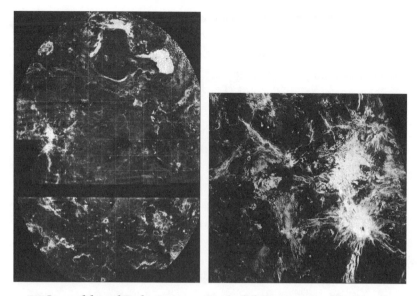

5.1 Ground-based Radar Images: *On the left is a mosaic of the "Earth-facing" hemisphere made at Arecibo in 1975 and 1977. Ishtar Terra is the pear-shaped area in the north. The bright area to the east is Maxwell Montes. Beta Regio is the bright region in the central western region. On the right is a much more detailed mosaic of the Beta Regio area made at Arecibo in 1989. (D. Campbell)*

in the nineteen-month interim. Thus, ground-based imaging efforts assumed the same rhythm, constrained by orbital dynamics, as the spacecraft "race" to Venus. Images made in 1972 had five times better resolution than those obtained in 1970, and allowed the first attempts to identify circular features that might be impact craters. If we could tell the number of craters, we would have some idea of the age of the surface—a very important clue to the planet's history.

In 1975, a map of the mid-northern latitudes of the Earth-facing hemisphere made from Arecibo revealed a very broad, radar-dark (and thus presumably flat) area. To its east was a fantastically bright (and presumably rough) feature, which was named Maxwell (after the Scottish physicist who pioneered the theory of electromagnetism in the 1800s, eventually making radar possible). We later identified these features on the *Pioneer* topography map (made in 1979) as a high, broad, and flat plateau (now called Lakshmi Planum), and a huge and very rough mountain range (Maxwell Montes).

The quality and coverage of ground-based radar images of Venus gradually improved, and we pored over every detail of these images as if they were secret treasure maps. They were a valuable complement to the *Pioneer* topographical mapping, and they hinted at the variety of surface features to be discovered later. But they provided only incomplete, fuzzy, and surreal portraits of one hemisphere of the planet.

This restriction presented dangers that we had become acquainted with through our explorations of the two other nearest worlds, Mars and the Moon. Each of these planets is two-faced, defying attempts at global extrapolation from pictures of limited areas. As you can see on most clear nights, the near side of the Moon has numerous dark areas covered with basaltic lava flows.* It turns out these dark, smooth areas are almost entirely absent from the far side. Our first fly-by missions to Mars photographed only ancient and crater-filled portions of the surface, and the early verdict was that it was all as old and dead as the Moon. Then, in 1971 *Mariner 9* orbited and photographed the whole planet, revealing giant volcanoes, vast canyons, fields of sand dunes, and even ancient riverbeds. It showed us a geographically schizoid planet, with ancient, heavily cratered terrain in the southern hemisphere and younger, smoother volcanic plains in the north. Forcefully demonstrating the dangers of extrapolating a planet's nature from partial views, these revelations made us yearn even more strongly for a global radar mapping of Venus.

We could achieve that only by putting a radio telescope on an orbiting spacecraft.

THE BIRTH OF MAGELLAN

Magellan was a long time coming, but decades of papers, conferences, and controversies became moot the moment its images arrived on Earth. In one of my first scientific jobs, in the late '70s and early '80s, I was an undergraduate research assistant at Brown University helping to find impact craters on the surface of Venus. If the surface was young and active, constantly recycling itself like Earth's, there should not be

---

*These form the shapes we North Americans call the "man in the Moon" and our southern neighbors call the "rabbit in the Moon."

many craters. (Over 60 percent of Earth's surface has been created by plate tectonics in just the last 200 million years.)

We searched all the available images and maps, identifying circular features and trying to rule out other possible origins for them (volcanic, for example). Basically, we looked for circular features with smooth floors and rough, raised rims. They also needed to have the right *depth-to-diameter ratio* (as defined by the well-known shapes of impact craters on the Moon and other planets and on our ideas of how the Venusian environment would affect these shapes).*

For some reason, perhaps because I liked the names, two surface features stand out in my memory of those days: Colette and Sacajawea. They are both circular depressions, about one hundred miles across, on the large elevated plateau (Lakshmi Planum), which forms the western part of the continent-sized northern highland of Ishtar. Both looked like impact craters, but we did not have a great sense of certainty about any of our conclusions then. The view was too fuzzy.

We knew that if things went well within our lifetimes (and they have!), spacecraft-borne radar might produce images that would clearly distinguish craters from noncraters, but that didn't stop us. Like impatient children who just can't wait for the toy store to open, peering through dim windows in hopes of a glimpse of the genuine articles, we savored every clue, knowing full well that an upcoming spacecraft radar mission might turn on the lights and make our efforts obsolete.

In those days we were all excited about a mission that was supposed to fly in the late 1980s: the *Venus Orbiting Imaging Radar*, or *VOIR* (French for "to see"). The centerpiece of *VOIR* was to be radar imaging of the surface, but it would also have had great experiments for further study of the atmosphere. It was to be an ambitious and comprehensive mission, loaded with sophisticated instruments like the *Vikings* that went to Mars and the *Voyagers* that explored the outer solar system during planetary exploration's golden age. Like these other missions, *VOIR* was designed during the 1970s, a time when NASA was still riding high on the wave of confidence and funding that accompanied

---

*We expected, with the high surface temperature, that large craters would have lower rims than on the Moon, because the rock would flow more easily over geologic time. We expected the thick atmosphere to prevent material from being thrown to great distances during a cratering event, so the rough ejected material should be piled up near the rim. We also considered the strength of the surface gravity.

the *Apollo* Moon missions. *Viking* and *Voyager* surfed safely away from Earth before this wave broke and began to recede. *VOIR*, which was to be the next great step in our exploration of Venus, got dragged down in the undertow. Its budget had swollen to over $800 million while support for space science was steadily falling. In the 1981 congressional markup of President Reagan's 1982 budget, all funds for *VOIR* were deleted. It was dead, Jim.

The cancellation of *VOIR* was the "unkindest cut of all" for the Venus science community, and horrible for the spirit of planetary science in general. It felt like a promise broken, a voyage we had been psyching ourselves up for, which we were now told we could not take.

But a great idea is hard to kill. A more modest mission ($270 million) with only radar imaging and an altimeter was resurrected in 1983 and called *Venus Radar Mapper.** This was a fairly uninspired name for a bold mission of exploration, so in 1986 it was renamed after Ferdinand Magellan, the Portuguese-born explorer who is often said to have been the first person to circumnavigate the world.[†]

Magellan was an economy version of *VOIR*, but even this stripped down concept would be hard to pull off within the budget allotted by Congress. So, like cars in Cuba where money and parts are hard to come by, *Magellan* was resourcefully put together with spares and leftovers. And like the numerous families in Havana occupying the mansions of the privileged from a previous age, *Magellan* engineers scavenged the storehouses of the golden age to piece together a spacecraft. Some of the most important components were leftovers from the *Voyager* program. Its main antenna, rocket thrusters, and "equipment bus" (the octagonal framework to which all other parts were attached) were all *Voyager* spares. Its computers were extra ones from *Galileo*, and

---

*Further savings could have been achieved by also cutting the radar and just sending it to Venus without any instruments, but I don't think this was seriously considered.

[†]In reality, Magellan died before completing his mission. Of the five ships and hundreds of men who left Spain on Magellan's expedition in 1519, only one ship with eighteen men returned in 1522. Magellan himself was killed in battle in 1521 by "natives" in the Philippines whom he was trying to convert to Christianity. The scientists who named a mission in his honor 460 years later had better luck at living to see their dream come true: the dream of circling Venus with orbiting radar.

various components were plundered from other missions. Against all odds, it started to look as if *Magellan* could be on time and on budget.

But then a disaster occurred. *Magellan* suffered a major delay and further budget increases from an accident that no one could have predicted, except maybe some engineers at Morton Thiokol to whom no one wanted to listen: on the freezing Florida morning of January 28, 1986, the space shuttle *Challenger* exploded seventy-three seconds after launch, killing schoolteacher Christa McAuliffe and six other brave explorers. NASA shelved all shuttle launches for thirty-two months while the accident was investigated. Among the victims of this delay was *Magellan,* which had been scheduled for launch on board the shuttle in May 1988.

Magellan was given an October 1989 launch date, aboard the refurbished shuttle, but there was a further complication. The *Galileo* mission to Jupiter (by way of Venus) had also been bumped from its anticipated launch date by the *Challenger* disaster. The *Galileo* team, which had suffered more than its share of delays starting before *Magellan* was a gleam in anyone's eye, had recalculated possible trajectories for new launch dates. Because of a complicated path to Jupiter, making one fly-by of Venus and two of Earth to pick up gravitational steam, there was only so much flexibility in *Galileo's* schedule; Earth, Venus, and Jupiter all had to be in the right place at the right time. The planetary configurations were not subject to negotiation, and NASA determined that *Galileo* needed the October '89 shuttle launch. *Magellan* got bumped.

It actually got bumped to an *earlier* launch date, but this meant that it would arrive at Venus later! At the time of the new launch date in May 1989, Venus and Earth were not lined up in the right way for the originally planned four-month trip to Venus. *Magellan* would have to spend fifteen months in space, orbiting the Sun one-and-a-half times before reaching Venus. This was not at all optimal, because a longer spaceflight allows more time for something to go wrong. But we were grateful to get *Magellan* on its way. At least once in space it could not be canceled, or so we thought. Later, the *Magellan* team had to fight repeated threats to cancel the mission while the spacecraft was successfully operating at Venus.

Magellan left Earth aboard the space shuttle *Atlantis* on May 4, 1989. This was NASA's first launch of a planetary spacecraft since

Pioneer Venus in 1978 and the first ever on board the shuttle. Some problems arose on the fifteen-month trip to Venus: for unknown reasons the spacecraft started overheating on a few occasions while in interplanetary space. We solved this problem by rotating the spacecraft and using the large main antenna like a beach umbrella to shade and cool it. At one point the star scanners, which constantly checked the position of some bright stars to keep the craft oriented, got confused and didn't know where they were looking. One gyro went screwy and was retired, but there were several backups on board. There were also the requisite computer glitches, but for the most part these troubles were of the sort that have become almost routine for interplanetary flight, and on the whole the trip across space was relatively, mercifully mundane.

AROUND THE WORLD IN 243 DAYS

We've come a long way, we've come a long way and we never even left L.A.
—MICHELLE SHOCKED

Magellan is basically just a big, solar-powered radio dish. The plan was to go into an elliptical orbit over the poles of Venus, with a period of just over three hours.* During the low part of each orbit, when *Magellan* would be less than two hundred miles above the surface, the twelve-foot-diameter main antenna would be pointed toward the planet, flashing radar pulses at the surface and catching them on the rebound. For this thirty-seven-minute segment, starting at the north pole and swooping southward, *Magellan* would map one thin strip of the surface. Then during the rest of each orbit, which stretched to more than five thousand miles from Venus, the spacecraft would turn around so that the big antenna would face Earth and "download," sending its latest data-catch homeward. Partway through each of these playback periods, *Magellan* would take time out to check its orientation against familiar stars. Then it would finish transmitting to Earth and, approaching the low part of its orbit again, turn back toward Venus to image the next swath of ground (see Figure 5.2).

These would be the longest and thinnest pictures ever taken: each orbit would produce a "noodle," an image fifteen miles wide and ten thousand miles long, running north to south. Consecutive orbits

---

*By *period* we mean the time to complete one orbit. Earth's orbital period is one year.

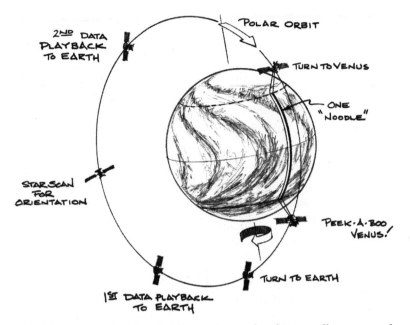

5.2 **Magellan's Mapping Orbit:** *During each orbit* Magellan *swooped low over the planet and pointed its main antenna downward for thirty-seven minutes, imaging a "noodle" of the planet; then it pointed toward Earth and sent down the new catch of images. (Carter Emmart)*

would capture adjacent noodles, with a slight overlap. The idea was to put *Magellan* into an orbit that stayed in a fixed position, and let Venus slowly turn beneath it, so that the entire surface could be viewed in 243 days, as Venus made one complete rotation.*

The precious signals to and from *Magellan* would be sent and received on Earth with the Deep Space Network, a global communications system with large radio dishes at Goldstone, California; Madrid, Spain; and Canberra, Australia. Such a wide geographical distribution is necessary to keep in constant touch with distant spacecraft as Earth turns. The Deep Space Network would forward *Magellan's* calls to

---

*This is Venus's rotation rate in an "inertial" frame, or relative to the "fixed distant stars," as opposed to its 117 (Earth) day solar day, which is affected by its orbital period. I know this is a bit confusing, but that's the spinning, turning universe we live in, and if we want to understand and navigate it we just have to get used to it.

NASA's Jet Propulsion Laboratory (JPL), in Pasadena, California, northeast of L.A., where computers would assemble the stream of digital data into images.

JPL is a collection of modern, efficiently built buildings ringed by a barbed-wire security fence in the foothills of the San Gabriel Mountains, a place that on some August days is so hot and smothered in Sun-dimming smog that it bears an uncomfortable resemblance to the surface of our sister planet.* It is a somewhat formal institution (one of the few places where space scientists can sometimes be found wearing suits), resembling a college campus where they forgot to admit any students. It is also the place where some of the most wonderful magic of our time has been wrought. As *Magellan* neared Venus, more magic was in store if everything worked.

Magellan finally made it to Venus on August 10, 1990. The first thing it had to do was fire a large rocket motor to slow down enough to enter into orbit. If this didn't work, the mission would be a very brief and sorry fly-by that would return no images. The well-funded *Vikings* and *Voyagers* were made in pairs to guard against catastrophic failure, but there was only one *Magellan*, and there would be only one chance. Like *Pioneer* twelve years earlier, *Magellan* had to make this move on its own, behind Venus and cut off from direct communication with Earth. The ground controllers were perhaps a bit less anxious this time, since we had gained some experience and confidence with another decade of space missions. But it was by no means a sure thing.

As the first critical moment approached, ground engineers hoped they would hear nothing at all from *Magellan*. An early signal would mean that *Magellan* had not burned its rocket, had not slowed down and entered orbit, and was now flying off in a useless orbit around the Sun. That moment came and went, with sweaty palms followed by a big sigh of relief. On the other side of Venus, *Magellan* had fired its big solid-fuel rocket for eighty-three seconds, sending the craft into an elliptical orbit around Venus with a period of three hours and twenty-six minutes—very close to the mission plan. A few minutes later the crucial signal saying "I'm here and everything's fine!" was received at JPL to much screaming and applause. This is one of the most satisfying

---

*Venus is much hotter, but at least it's a dry heat.

moments in planetary exploration—the engineer's equivalent to making the winning three-point jump shot in the finals.

MAGELLAN, *PHONE HOME!*

Let my machine talk to me. Let my machine talk to me.
—R.E.M., "World Leader Pretend"

Five days later we turned on and tested *Magellan*'s radar system. It checked out well, so the next day, August 16, it was time to unveil some "noodles" of Venus and see what was really down there—a trial run of *Magellan*'s imaging systems to prepare for the systematic mapping of the planet.

Magellan had some hairy moments early on, to put it mildly. Can you imagine what it was like for the people who had poured major chunks of their careers and souls into an interplanetary craft to receive word that the signal had been lost, that there was an unexplained, potentially fatal glitch?

Everything went fine at first. *Magellan* was sending down what seemed like good data, although it would take a few hours for the JPL computers to turn them into images. Then, after one and a half orbits of mapping, the signal abruptly stopped. New commands were urgently sent up, but *Magellan* was completely silent. An emotionally strange moment came when the first one and a half noodles were finally processed into images. The pictures looked good, and showed teasing hints of slices of impact craters, volcanic flows, and ridges. Ordinarily this would have been a time for rejoicing, but now their spacecraft was maintaining total radio silence and no one knew what was wrong. Would this one tiny, thin strip of surface map turn out to be the entire legacy of all the years of dreaming, hard work, and hundreds of millions of dollars? This would have made the fate of the *Magellan* mission a bit more like that of its namesake.

The nightmare scenario seemed all too possible, so no one was in the mood to celebrate the first image. This was only two months after NASA's mega-embarrassment at learning that the Hubble Space Telescope had been put into orbit with a serious design flaw that technicians should have caught on the ground. The Hubble problem was later fixed with a fantastic shuttle repair mission that amounted to the most expensive and daring optometry ever performed. But no repair mission would be possible if *Magellan* refused to phone home. The press has often made a bigger deal of NASA's occasional failures than its numerous successes. You could be sure that the total loss of *Magellan* would get a bigger headline in the *New York Times* than the complete mapping of Venus ever would, even though the latter is certainly a more significant event in human history. Spacecraft losses are sudden and catastrophic, and mapping a planet is a slow, somewhat methodical business, with gradual revelation that doesn't make for tidy news and sound bites.

Presumably, *Magellan* had somehow become disoriented and could not find the direction to Earth. If so, it should have entered a preprogrammed "safe" mode, in which it methodically turns and searches for familiar stars to reorient itself. Hours went by and there was still no word from the craft. If this was a minor disorientation, it should have gotten back in touch by now. Was *Magellan* tumbling out of control? After fourteen nail-biting hours, we received a signal from a still-intact, although somewhat confused, *Magellan*. For mysterious reasons, it had been staring into space, with its computers strangely and seemingly irrationally switching between various emergency routines.

Years earlier NASA had run its first astronaut candidates through extensive batteries of psychological tests to ensure that they had "the right stuff," so the strangeness and enormity of their situation when they found themselves in orbit wouldn't cause them to flip out, leaving them unable to perform their technical routines. No one expected this unstable behavior in a robot spacecraft. This incident came to be known as *Magellan's* "walkabout."

We had regained contact and the spacecraft seemed okay. But five days later, while engineers were still trying to pinpoint the problem, *Magellan* went astray again. This second walkabout lasted longer and seemed potentially more serious. After almost eighteen hours of agony, the spacecraft signal was found once again and *Magellan* started to behave itself. These troubling incidents shook up the *Magellan* team quite a bit. Since no one understood them, we did not know how likely a recurrence was.

Several weeks went by with no repeat of the problem, so we decided to start mapping the planet and hope for the best.

Magellan began mapping on September 15, 1990, after a month's delay.* Slowly, noodle by noodle, Venus was revealed, but further problems arose. The disturbing walkabout problem recurred three times at seemingly random intervals during the first 243-day mapping cycle, once for thirty-two hours. Each incident left a gap in the global map. Then, in February 1991, the spacecraft began to overheat dangerously. To counter this, we cut short the 37 minute mapping segment of each orbit by ten minutes, so the main antenna could be pointed at the Sun and shade the spacecraft. It worked, but we lost ground coverage. In May the overheating worsened considerably and controllers had to give up 55 percent of each mapping segment. Still, when cycle 1 ended on May 15, *Magellan* had passed close over the entire surface and captured images of 84 percent of it. Not bad for starters.

During the second tour, cycle 2, the spacecraft lost a transmitter, and problems in the backup transmitter required cutting the signal rate in half. But the cause of those damn walkabout episodes was finally discovered in July 1991. An obscure computer routine in the system that controlled the spacecraft's orientation, given certain rare combinations of inputs, had gone into a runaway state where it would babble to itself for hours. Once found, the problem was easily corrected. The gaps were filled, and by the end of cycle 2 in January 1992, *Magellan* had imaged 97 percent of the planet.

PEELING BACK THE CLOUDS

Finally, the fog was lifting from the surface of Venus. It was not a gradual clearing but a sequential peeling back of the clouds in thin strips from west to east. The *Magellan* images were so clear that we felt we were seeing Venus for the first time. A certain psychological and visual threshold was crossed. We humans are visual creatures and surface dwellers, so we need pictures of a planet's surface to identify it as an environment to be compared with other places we know. It was like finally, after decades of squinting, getting the right prescription for

---

*This was the beginning of "Cycle 1." The long-term plan and the requests for continuing funding were split into 243-day mapping cycles, during which the whole planet spun once beneath *Magellan*'s gaze.

your glasses and seeing the world in focus. Only after *Magellan* did Venus seem like a real place, with landscapes that we could compare on an intuitive level with our home planet. Thus begins anew the task of trying to explain what is going on there. For students of planetary evolution this is where the fun, and the work, begin.

Before we could even begin to address the "big-picture" questions of Venusian long-term history and evolution, there was a lot of information to sort out. *Magellan* has suddenly presented the planetary science community with an embarrassment of riches equal, in digital terms, to more than twice the data from all previous planetary missions combined! Thanks to *Magellan,* we now have global maps of Venus that are in many respects superior to those we have of Earth, since most of Earth is under the oceans.

Even though I am writing this five years after the end of *Magellan's* first mapping cycle, we are just beginning to figure out what it all means. Our understanding of Venus is a work in progress, so much of what I will tell you is still highly uncertain and contentious. The data are still new and the ideas raw. Scientifically, Venus is a frontier planet, now more than ever. Like a planetary atmosphere far from equilibrium, the air at our meetings is thick with potentially volatile combinations of competing ideas and explanations. As these concepts meet, they generate sparks, heat, and, we hope, light.

How would you describe Earth to someone who had never seen it? Which places and details would you choose to mention? Venus has three times the land area of Earth! I can't possibly describe the whole planet, but I will do my best to give you a feel for its character with a few illustrative place descriptions. You can find many of the names I mention on the *Magellan* altimetry map in the color insert.

VOLCANO WORLD

So what do we see, through *Magellan's* radar eyes, on Venus? We see volcanoes. When Galileo publicized his observations suggesting that the universe was so much vaster than the small province of Earth, the Catholic honchos felt that the turf of their God was somehow being challenged (rather than enlarged). Clearly some gods are much more flexible.* One look at almost any *Magellan* picture shows that the

---

*More Catholic, one might say.

domain of Pele, the Hawaiian volcano God, extends well beyond Earth. Venus might just as well be called Volcano World. It is unparalleled anywhere in the solar system for the sheer number, astounding variety, and pristine state of preservation of its volcanic features.* Volcanic landforms occupy almost 90 percent of the surface, with styles ranging from the strikingly familiar to the downright bizarre.

Vast low-lying areas are covered by flat, relatively featureless flows of lava that may resemble the Columbia River flood basalts of western North America or the dark, smooth Mare regions of the Moon. These are blue areas of the map in the color insert. These plains are dotted with thousands of small (less than twelve miles across) "shield" volcanoes (also common on Earth). Their shield-like shapes have been built up gradually by repeated flows of runny lava. Outbreaks of these small volcanoes seem to crop up everywhere across the face of Venus, like pimples on a teenager. Large clusters of them, typically one hundred miles across, are gathered in certain areas called "shield fields." Probably hundreds of thousands of the small shields are distributed around the global plains. We find a similar number of small volcanoes on Earth's ocean floors (see Figure 5.3).

A smaller number of gargantuan Hawaiian-style shield volcanoes, hundreds of miles across, tower up to five miles above the surrounding plains. Their gently sloping flanks are covered by numerous overlapping lava flows, the most recent of which appear as if they happened yesterday. In the *Magellan* images they are indistinguishable from active volcanoes. We tried and failed to see the "smoking gun" of a volcanic flow changing position between *Magellan* mapping cycles, but an identical test on Earth would probably not reveal volcanic activity either.

In the pictures of shield volcanoes in the black and white insert (and on the cover of this book), you can see evidence of many episodes of flow, indicating a long history of activity. The brightest flows are the roughest, and generally overlie darker ones, so they are more recent (see Figure 5.4). Magellan's improved vision revealed that many of the large highlands in the old *Pioneer* topo map are capped by giant shield volcanoes. Since they are usually located on top of broader topographic rises, the magma that flows from these volcanoes probably comes from huge masses of hot, upwelling material (hot spots) beneath the lithosphere.

---

*Although Jupiter's active moon Io is no volcanic slouch either.

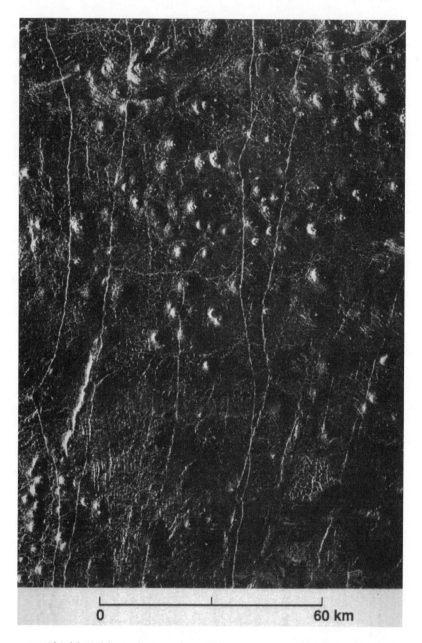

0   60 km

5.3 Shield Field: *A cluster of small volcanoes, each about two miles across. (NASA)*

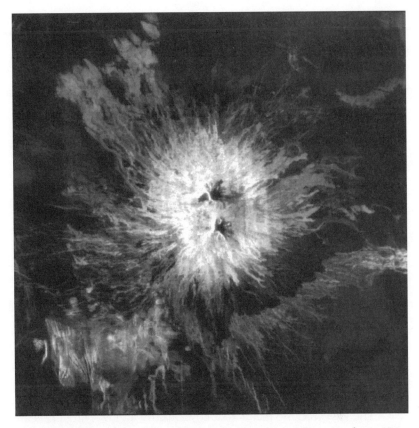

5.4 Hawaiian Style: *This shield volcano, Sapas Mons, is about 250 miles across. Numerous overlapping lava flows can be seen on its flanks.*

Clusters of smaller, flat-topped "pancake domes" with steep sides are sites where more viscous lavas have squeezed up through the crust. (See Figure 5.5 and insert.) We have identified more than 150 of these features, which have no obvious close terrestrial analogy. There are domes of similar shape on Earth,* but the pancakes on Venus are ten to one hundred times larger—six to forty miles across. Whatever these pancakes are, they are recognizable as volcanoes, and the qualities of the lava—which hints at its chemical makeup—can be inferred from the way it has flowed. Any clues about what the rocks are actually made of, and how this varies from place to place, are tremendously valuable.

---

*Called *rhyolite domes,* after the type of high-viscosity rock, similar to granite, which forms them. A cluster of these domes is found in the Mammoth Lakes area of California.

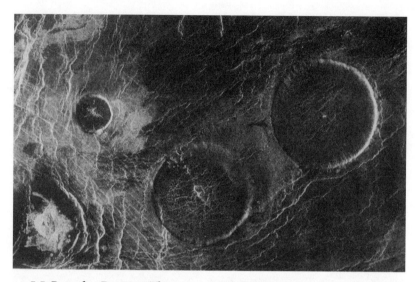

5.5 Pancake Domes: *These steep-sided volcanoes are each about fifteen miles across and twenty-five hundred feet high—a challenging day's hike.*

By examining the global distribution of different types of volcanoes, we can look for signs that the crust of Venus, like Earth's, is differentiated into areas of differing rock types.

The vast volcanic plains that cover nearly all low-lying areas are the long-sought global "oceans" on the surface of Venus—frozen oceans of basalt. One of the most astounding surface forms discovered by *Magellan* furthers the analogy: this ocean is fed by rivers! We see numerous thin, meandering channels, typically a mile wide and up to thousands of miles in length. The longest of these, Baltis Vallis, winds more than forty-two hundred miles over the plains.* Baltis is longer than the Nile and thus can safely be called the longest river anywhere within several light-years of here. (See *Magellan* picture in black and white insert.) On Earth or Mars we would interpret such features as evidence of past or present running water. The analogy with structures carved by terrestrial water goes quite deep. We see fanlike river deltas, meanders, and bars, and places where streams have flooded their

---

*It could be much longer. Both ends are covered by younger flows, so we don't know the original length. Baltis is an ancient Arabian goddess identified with the planet Venus.

banks. There is evidence of channel and flood-plain migration of the kind that usually implies a "mature" stream with a long history of flow, erosion, ebb, and flood. On Venus, where it is far too hot for liquid water, perhaps a stream of lava with the right chemical composition could flow like a river. We have had to use a lot of imaginative physics and chemistry to come up with a suitable model (see Figure 5.6).

Whatever this stuff is, it has to stay molten for a long time to flow so far and carve rivers into the surrounding rock. Some sulfur-rich rock types might have the right properties. Another candidate is a carbon-rich rock called carbonatite. Carbonatite volcanism is rare on Earth, but it does exist. There is a carbonatite volcano in Tanzania where the lava has a melting temperature of 914 degrees Fahrenheit, pretty close to room temperature on Venus. Some scientists have even speculated that there might be a carbonatite "aquifer" below the surface of Venus, like Earth's ubiquitous underground water—a liquid reservoir that can erupt into surface flows under the right conditions of heat and pressure.

We have had to go back to basics, look at the fundamental properties of flowing fluids, and ask deep questions about rivers on Earth: what causes river meanders of a certain size and type, and how are

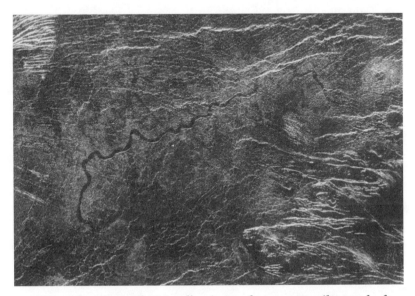

5.6 Venusian River: *This* Magellan *image shows a 125-mile stretch of a sinuous "river" channel winding across the volcanic plains.*

they related to the physical properties and flow rate of the liquid that carves them? Topographical analysis of the flow paths produced a puzzling result: some of them seem to flow uphill! Since gravity is gravity on any planet, the ground in these places must have shifted since the time the rivers ran. The Venusian channels are a delightful find, and they make a textbook case of the way comparative planetology forces us to dig deep for physical understanding as we struggle to explain how unfamiliar materials can exhibit familiar behavior in alien environments.

The converse of this is also true in abundance: familiar materials can take on new properties and unforeseen behavior in response to a very different set of physical conditions. On Venus we see many kinds of features that seem to be volcanic but that do not fit so conveniently into our classification and naming schemes derived from terrestrial geology. We have given names to some of these, inspired by the terrestrial fauna their appearance evokes: our new maps of Venus are populated with *ticks, arachnoids,* and *anemonae.* The names are deliberately nongenetic: when we're not sure what something is, we try to pick a neutral name that does not imply an origin theory, which would bias our investigations.

Still, for many of these features, we do think we know what they are. Some of them resemble modified terrestrial volcanoes. Ticks are volcanoes with flanks scalloped out by numerous huge landslides that have left ridges resembling insect legs jutting out (see Figure 5.7). The Venusian ticks resemble some seamounts, volcanoes sculpted by undersea landslides on the bottom of Earth's oceans. Arachnoids are volcanic domes surrounded by spiderweb-like patterns of fractures and ridges. Anemonae are volcanoes with petal-like lava flows extending outward onto the surrounding plains. (See *Magellan* pictures in black and white insert.)

LIKE NOTHING ON EARTH

One major type of volcanic feature found on Venus has no terrestrial counterpart. The giant circular features that we call *coronae** are a

---

*This is plural. The singular is *corona.*

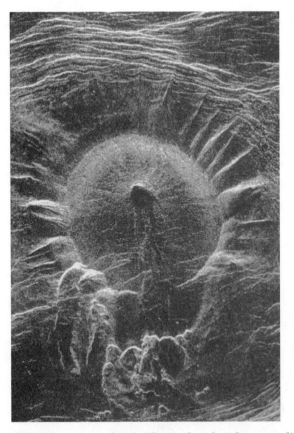

5.7 The Tick: *This is a volcanic dome that has been modified by landslides, leaving "legs" of ridges in between. The feature is about seventy miles across.*

uniquely Venusian landform, at least in this solar system.* A typical corona is surrounded by a concentric ring of ridges and fractures, and just outside of this is usually a moat or trough. Planetary geologists have classified many kinds of features on Venus, all different in detail, as coronae. Some coronae have central mounds; others are more craterlike, with low depressions inside. Some have a multitude of small shield volcanoes inside them, and some have huge flows of lava spreading out from their perimeters. What makes them all coronae is that they all have the circular ring of fractures. The exact number is hard to

*Unless we find them on Pluto or the unmapped parts of Mercury (unlikely). Some people claim they've seen coronae on Mars, but others dispute this.

determine, because not everyone can agree on what is or is not a corona, but we have mapped hundreds of them at widely distributed locations around the globe. (See Figure 5.8 and black and white insert.)

Although most coronae are one hundred to two hundred miles across, some are as small as fifty miles, and there are a handful of giants, including one with a diameter of more than fifteen hundred miles. This mighty mother of all coronae is named, appropriately enough, *Artemis Corona.** Most coronae are named after fertility goddesses, although, as befits the wide range of physical types, there are a few exceptions, including coronae named after Amelia Earhart and Florence Nightingale.

We learned about coronae for the first time from the earlier, more limited *Venera 15/16* radar mapping in the mid-1980s. Some investiga-

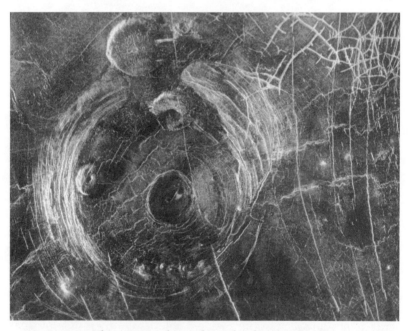

5.8 Corona: *This corona, located south of Aphrodite Terra, is 120 miles in diameter. Pancake domes can be seen within and adjacent to the corona.*

*Artemis is an Olympian goddess of the moon and of the hunt, and a protector of women. She was earlier identified with the Roman Diana, and still earlier with the great mother of nature.

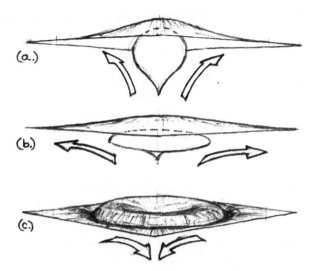

5.9 Formation of a Corona: *This cartoon shows possible stages in the growth of a corona: (a) A rising blob of bouyant rock pushes the surface up into a dome. (b) It hits the lithosphere, spreads out, and flattens. (c) The structure cools and sinks, leaving a circular ring. (Carter Emmart)*

tors proposed that they were the vestiges of giant, ancient impact craters left over from the intense bombardment that blasted the whole solar system soon after planet formation and ended about 4 billion years ago. The oldest areas of the Moon, Mercury, and Mars date back to this distant epoch. If this interpretation of coronae were correct it would mean that large areas of the surface of Venus were extremely ancient. It would argue for an inactive and dead world nothing like Earth. But planetary geologists have become very good at identifying the telltale signs of impact craters from orbital imagery and ruling out the imposters. Most *Venera* scientists doubted that these strange features were impact craters, and with the improved *Magellan* imagery it is now obvious that they are not.

Coronae seemed truly alien to terrestrial geologists. Was the term just a convenient catchall classification scheme for large circular features with no terrestrial analog? Or was there a deeper connection, some behavior and history common to all coronae? We now think that their diversity can be explained by placing them in a developmental sequence, with the different forms representing coronae caught in the act at different stages of an overall life history. First blobs of hot rock and magma squeeze up under the crust, forming blisterlike domes. They then spread out under their own weight as the rock "relaxes" like an overambitious sandcastle. As it sags outward, the ground cracks, forming the circle of ridges. Then finally, as the underlying rock cools, the ground subsides, leaving an internal depression (see Figure 5.9). We find a wide variety of coronae because we see them in different stages of this sequence. Some are still in the process of squeezing up, some blisters have "burst" recently, bleeding lava onto the surface, and some long ago ceased to be active and have had time to subside.

Coronae are found only in certain areas of Venus, probably where the lithosphere is thin enough to allow magma to push the surface up into domes. If so, then their global distribution may indicate large-scale variations in the thickness of the lithosphere.

It is a common practice for people fishing for bluefish off the coast of New England to search the horizon for flocks of terns. These birds feed on silversides, small fish driven to the surface by schools of bluefish. The fishermen have learned that where there is a flock of birds there is almost always a school of blues beneath the surface in a feeding frenzy. When you see birds, you race your boat over and cast into the swarm, and you are likely to land a bluefish. Planetary geologists search for similar reliable links between things we see aboveground and things going on below. Perhaps we can learn to measure the thickness of the lithosphere in various places by studying the global distribution of coronae, and use the result to map the planet's subsurface structure.

Why aren't there coronae on Earth? We, too, have rising blobs of hot rock pushing up under the lithosphere, which result in many kinds of landforms. What could be different? The most common explanation given is lithospheric thickness. Just as the appearance of coronae in certain areas on Venus may indicate a thinner lithosphere in these locales, their absence on Earth could signify a global difference: Venus may be thinner-skinned than Earth, at least in the spots where coronae form, allowing a certain kind of rising interior plume to create structures we don't find on Earth.* The question of how thick the Venusian litho-

---

*Some scientists have suggested that Earth may have had coronae in the distant past, when its interior was hotter and its lithosphere thinner.

sphere is, and how it compares with Earth's, is important for theories of global tectonics, and we will discuss it further in that context.

Another possibility is that the absence of coronae on Earth is an illusion.* Much of the geology of our planet lies buried, covered by water and blankets of sediment—sand and silt carried by wind and rain and deposited in low-lying areas. Maybe Earth has coronae and we don't know about them. That I can even make such a statement is testament to a remarkable fact: in some ways we now know Venus better than we do our own planet.

ALTERNATIVE ROCK

The coronae are perhaps the best example of a common geological form on Venus that has no counterpart on Earth. Most large formations there are easily recognizable as variations on familiar themes. I think we were largely prepared for a more alien and unrecognizable place than the one revealed by *Magellan.* All our previous views of the Venusian surface had been distorted or fuzzy: a few Soviet lander images taken at low angles to the surface; (see black and white insert) vague, ghostlike ground based radar images of one hemisphere; the low-resolution *Pioneer* topo map; and the scientifically valuable but visually grainy *Venera* radar maps of parts of the northern hemisphere. When compared with the decent photographic images we had of all other planets,[†] this mishmash of obscure imagery encouraged the general impression that the surface of Venus was an odd place, utterly unlike Earth, that might be largely inaccessible to understanding by terrestrial analogy. These images, and decades of description of Venus as hostile and hellish, may have contributed to expectations of a weirder world than the one we found.

Instead, in many places Venus looks much like Earth. The likeness should not have been surprising, given all the other similarities between the two planets, and the tendency of materials in nature to

---

*Last night as I wandered around Denver's *Cinco de Mayo* celebration I came upon a twenty foot high promotional display for a Mexican beer company that proved there is at least one giant Corona on Earth.

[†]Except poor little Pluto, if you consider that a planet. See footnote on page 126.

repeat the same forms in different settings. Many of the same forces and processes are present, but under different conditions and maybe using different materials, so we would expect some combination of the familiar and the unexpected.

What we are really talking about here is the behavior of rock, on large and small scales.* There are several reasons why rock might act differently on Venus than it does on Earth. Most obviously, the enormously high temperature should make rocks less viscous (more "squishy," more likely to flow under pressure). But the dryness of Venus has the opposite effect. All rocks on Earth (even in desert areas) have some water inside, locked up in their crystal structure or trapped in their pores. The water is an internal lubricant that lowers their viscosity. It is hard to simulate on Earth rocks as dry as the ones on Venus, but we have done some experiments, and it appears that the rocks become much stiffer. The effects of high temperature and the dryness in Venusian rocks might cancel out to some extent, but we don't know exactly how much.

Venusian rocks might also behave differently because they are made out of different materials from those on Earth. The *Venera* and *Vega* landers performed chemical experiments, all of which indicated rocks similar to our familiar dark volcanic basalts. Almost all of the Soviet landings were confined to one small area of the surface, around the flanks of the broad rise we still call Beta, a designation from the earliest days of radar imaging. Although this is anything but an adequate sample of the surface, the dominance of basalts is supported by the *Magellan* images of widespread volcanic features closely resembling terrestrial basalt flows. If it looks like basalt, flows like basalt, and (at least in a few lander locations) smells like basalt, then it probably is basalt.

At one *Venera* lander site, *Venera 8*, the chemical experiments indicated a distinctly different kind of rock that might more closely resemble granite, the less dense, more viscous rock that makes up most of Earth's continents. A Russian scientist later attempted to pinpoint and study the *Venera* landing sites in *Magellan* images. He found that

---

*You might think that rocks have only one behavior: they just sit there like rocks. But over geological timescales, rocks flow, crack, melt, and sag, and this is what shapes the surfaces of worlds.

Venera 8 may have landed on or near one of the steep-sided pancake domes. This fits nicely, because the pancakes (along with the "rivers") are one of the few Venusian volcanic features with shapes that probably could not have been made by basalts. They require something more viscous, like a granite.*

Even where Venusian rocks chemically resemble terrestrial ones, lavas erupting to the surface encounter a very different environment. We have already mentioned temperature and dryness, but there is also the high density and pressure of the air. You get colder faster under water than in air because the higher the density of a surrounding fluid, the faster it carries away heat by conduction. Thus, even though it's hotter there, Venusian lava flows might cool faster than flows on Earth. This is quite important because of the temperature control of viscosity, which can affect the length of volcanic flows and the shapes they assume. This is another area where the study of Venus has stretched our Earthbound science. Our models that predict how long lava will flow on Earth before turning to stone *should* work for Venus. But we have had to modify them to include all of the factors described here.

Some of the Venusian volcanic forms, especially ticks and pancake domes, resemble undersea features on Earth. This is interesting because the atmosphere of Venus resembles our oceans in certain ways. Lavas bursting to the surface on Venus encounter a resisting fluid that is denser and better able to conduct heat than our air, but less dense and less conducive to conduction than our water. Maybe it makes sense that volcanoes there should choose some forms found on our continents and some found under our seas.

READING THE CRACKS

Volcanism, while pervasive and varied, is not the whole story on Venus. Also present in abundance are a large number and variety of places where internal forces have pushed, pulled, warped, and squeezed the surface, forming giant cracks, valleys, cliffs, fractures, faults, and

---

*Earlier I compared the pancakes to terrestrial rhyolite domes. Rhyolite is basically the same thing as granite. When it solidifies from melt below ground it is called granite. When it flows on the surface, cools, and solidifies, it is called rhyolite.

folds—all features due to various forms of tectonics.* For the moment I will sidestep the question of whether Venus has an Earthlike global system of plate tectonics, because we need to hear more evidence. Much of that evidence consists of the forms and distribution of the tectonic structures that record the history of forces acting on the Venusian crust.

The second most widespread kind of terrain on Venus (after the volcanic plains) is also one of the strangest. About 10 percent of the planet, including most of the elevated regions, is composed of intensely deformed, very rugged areas we call *tesserae*† (see Figure 5.10). Some high plateaus consist of nothing but tessera,‡ huge tracts of cracked and wrinkled land extending for thousands of miles. The tesserae are really a mess. They appear to be veterans of many separate past episodes of tectonic deformation.

One large area of tessera is the giant plateau named Alpha Regio. This was the first spot on Venus identified in ground-based radar images because of its great radar brightness that results from its incredible roughness. Alpha Regio is about seven hundred miles across, the size of Texas and Oklahoma combined, and it stands more than a mile above the surrounding plains. It's easy to find on the map in the insert at about 20 degrees South and 0 degrees longitude. Most major highlands areas on Venus have some areas of tessera terrain and those that are not dominated by shield volcanoes are mostly composed of it. The two giant, continent-sized elevated regions, Ishtar in the north and Aphrodite on the equator, both have major areas of tessera.

The many layers of superimposed faults (cracks) in the tessera give it an unruly, chaotic appearance and indicate that it has been pushed (compressed) and pulled (extended) many times from different directions. In most places the compression seems to have come first and the extension later, but the tesserae are so jumbled that the historical sequence is often hard to unravel. One thing everyone agrees on is that the tesserae are the oldest places on the surface. We know this because they are always covered up at their edges by the surrounding volcanic

---

*My geology textbook says, "The word tectonic refers to the deformation of the rocks of the Earth's crust and the structure of the crust in general." I think we can get away with applying it to Venus here.

†This comes from a Russian word that means "tiled."

‡*Tessera* is single and *tesserae* is plural.

5.10 Tessera: *This portion of Alpha Regio, about eighty miles across, is composed of the intensely deformed terrain called "tessera." The complex structure is the signature of a long history of folding and faulting.*

plains. This tells us that the tesserae were there first. In fact, being oldest and found preferentially at high elevations, the tesserae seem like islands of an older surface that survived the volcanic flooding that covered most of the planet. Does this mean that there are tesserae everywhere, underneath the plains, with the pieces we can see representing just some small vestiges of what the whole surface once looked like? Maybe the intense history of deformation seen on the tesserae means that Venus was more tectonically active in the past than it is now; or maybe these areas are just showing their age, having been around for so long that their skin has become cracked and wrinkled.

What is the underlying cause of the large plateaus of tessera, and how do they reflect structures and motions in the planet's interior? Attempts to answer these questions have led to what is called the hot-spot–cold-spot controversy. Scientists have proposed two kinds of models: upwelling and downwelling. In upwelling models, the tesserae are the sites of ancient hotspots where abundant melting below the surface created low-density rocks that thickened the crust and made elevated terrain. In the downwelling model, tesserae were formed over places where colder material plunged into the planet's mantle. Why would this create an elevated area? As material is pulled from all directions toward the mantle downdraft, the converging flows collide, forcing the surface upward in the way colliding continents on Earth build up mountain ranges. The lower-density crustal scum piles up on the surface.

That these two essentially opposite concepts for tessera formation are still both in the arena of ideas should give you a sense of how new all this is. On the timescale that it takes to figure out a world, *Magellan* happened just yesterday, and we still have our work cut out for us. Whichever theory explains the original source of these great uplifts, it is obvious that areas of tessera later went through a long and complicated tectonic evolution. Like ancient manuscripts, cracked and torn, the tesserae are difficult to read, but they hold stories about the Venusian past that are preserved nowhere else on the planet.

The tesserae are the most confused, and at least among the most confusing, tectonically deformed areas on Venus. Most patterns of deformation are simpler and easier to read. We find important clues to more recent history in the more subtle tectonic patterns found in the least deformed areas of the planet: the widespread global volcanic plains.

Different areas of the plains contain textures and patterns that record the long-term history of forces affecting the crust. These tectonic fabrics were woven by pulses of stress that passed through the landscape and left their marks for us to decipher. Large areas of the plains contain rhythmic, repeating patterns of closely spaced parallel ridges and valleys. These "densely fractured plains" are a sign that the lithosphere has been stretched apart, causing mild deformation spread over very large areas (see Figure 5.11).

Other, younger areas of the plains are split by various types of ridges that reveal sites of past compression. Geologists, using experience from the thirty years of comparative planetology that preceded *Magellan*, have read the history of the plains in these stress patterns (see Figure 5.12).

These remnants of old stress and strain crisscrossing the mostly bland, featureless plains preserve memories of the forces from below that have tugged at large portions of the Venusian crust, and its responses to these pulls. But the real action, the dramatic tectonics, occurs where more distinct features rise up out of the plains. These range in size from small volcanoes up to mighty Ishtar and Aphrodite.

Some of the most striking, and apparently youngest, tectonic features are associated with the giant shield volcanoes. These generally

5.11 Fractured Plains: *This picture shows an area twenty by fifty miles across in Guinevere Planitia. The plains are covered by a grid of fractures that were caused by stretching of the crust.*

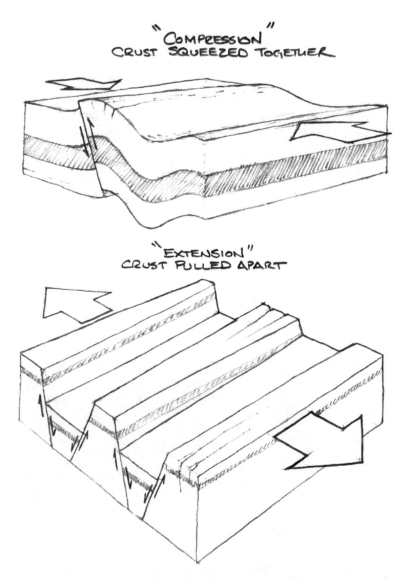

5.12 Compression and Extension: *Tectonic forces that compress and extend areas of the lithosphere leave their telltale signs on the surface. (Carter Emmart)*

occur in places where the surface has been domed upward, causing *rifts*, giant cracks rending the crust, extending outward from the volcanoes. It is as if the surface has been punched up from underneath by the material trying to get out of the volcanoes. This makes sense if the giant volcanoes are the sites of massive hot spots that have forced the lithosphere upward. Zones of rifting are also seen around the flanks of shield volcanoes on Earth and Mars.

Perhaps the most spectacular region of giant shield volcanoes and massive rifting is the area identified as Beta in the early ground-based radar images, which was also one of the first sights seen by *Magellan*. Each of *Magellan's* mapping cycles began in the vicinity of Beta and ended there 243 days later after the slow turning of the planet had swept the whole surface beneath the orbiting craft.

Let's take a brief, selective, circumnavigational tour of Venus, following *Magellan's* path from west to east, focusing on three of the most prominent and interesting highlands areas: Beta, Ishtar, and Aphrodite.

BETA: FLASHBACKS OF THE CRADLE

We are scatterlings of Africa on a journey to the stars.
—JOHNNY CLEGG

Beta Regio, fifteen hundred miles across, lies in the mid-northern latitudes at 280 degrees longitude. If you plunked it down on East Africa, it would cover Ethiopia, Kenya, and Somalia. The center of Beta is dominated by a colossal shield volcano, Theia Mons, which rises sixteen thousand feet above the plains and has lava flows running for hundreds of miles down its flanks. Theia Mons is located at the intersection of several giant rift valleys. The most impressive of these valleys, Devana Chasma, sixty to one hundred fifty miles wide and two to four miles deep, runs both northward and southward from Theia Mons for thousands of miles. Nearly a thousand miles to the north of Theia, this rift runs into Rhea Mons, a massive uplifted block of tessera that forms the northern part of Beta Regio.

Much controversy still surrounds the origin of Venusian highlands, but everyone seems to agree that Beta Regio is the site of a huge hot spot that is probably still rising from the Venusian mantle. Beta is probably currently active, both volcanically and tectonically.

Beta has always had a privileged position in our exploration of Venus. It was the site of our first ground-based radar images and the location of the first Soviet surface landings. *Magellan* began its journey of discovery there. And it is curiously reminiscent of a place on Earth that may have seen the beginning of another journey of considerable importance to us—the first ancestral stomping grounds of the human race.

The earliest human remains yet unearthed (approaching 3 million years old last I heard) have been found in the ancient lake beds of Tanzania's Olduvai Gorge, which is located within a larger geological structure known as the East African Rift System. Check out any map of Africa and you can see a network of giant cracks extending north to south through East Africa, threatening to split off the "horn." It starts in the north, by Djiboute on the Gulf of Aden. Running south from there is a line of lakes cutting through Ethiopia and Kenya into Lake Victoria. Farther south are two long, thin lakes: Lake Tanganyika, which forms the border between Zaire and Tanzania, and Lake Malawi. Still further south, the Zambezi River runs through Mozambique to the Indian Ocean.

These are some of the most obvious surface features of East Africa's great rift. Seismologists' maps of Earth's deep mantle show that a huge blob of hot material is rising under East Africa, uplifting, thinning, and splitting the lithosphere.* Africa is slowly coming apart along this rift, as the Somali plate pulls away to the east from the African plate, moving at a rate of one or two inches a year. The process has been going on for about 15 million years. That is not very long in geological time: all of human evolution, which may have begun here, has taken place while the rift opened slightly wider. In another 100 million years or so this rift could become a spreading center. Watch this site: future home of a midocean ridge.

There are intriguing similarities between Devana Chasma, the giant rift running through Beta Regio, and the East African Rift System. We may not have found a Garden of Eden on Venus (as some had hoped in the 1700s), but there is one area that reminds us of the old East African hometown of humankind. And the resemblance is not superficial: similar processes are going on above and below the ground at these two great hot spots on neighboring worlds. (See the black and white insert.) There are differences, too, and they may give us clues to the nature of rocks in the Venusian crust.

---

*Zones of rifting are generally zones of high heat flow, because whatever causes the rifting, in this case a mantle hot spot, usually brings heat with it. In areas of high heat flow the lithosphere is thinner, just as ice on a pond will be thinner over a hot water vent.

As a rift widens, stresses build up along faults between adjacent blocks of rock. Eventually the rock can no longer stand it: the ground gives way, and an Earth (or Venus) quake results. Devana Chasma is wider and deeper than the East African Rift and this leads geophysicists analyzing the forces along the sides of the two rifts to conclude that the Venusian rocks are tougher than their terrestrial counterparts. The great heights of the cliffs and the enormous size of the blocks of rock that have dropped into the rift suggest that stronger forces build up on Venus than on Earth before a break occurs along a fault. The reason might be the lack of water and corresponding strength of the rocks, possibly a hint that this effect is more important than softening by the high temperature. If this analysis of the difference between these two great rift valleys is correct, we may be seeing the effect of the incredible dryness of Venus allowing high cliffs, which would collapse on Earth, to stand.

ISHTAR: NO VISIBLE MEANS OF SUPPORT

As we follow *Magellan's* path eastward, our next stop is Ishtar Terra.* Ishtar was one of the places we had already studied a lot with pre-*Magellan* data. It lies far to the north, from just below 60 degrees to above 70 degrees of latitude on the "Earth-facing" side that we see in ground-based radar images. Ishtar's strange appearance in these images had sparked much speculation because there was nothing else like it anywhere on that hemisphere. We now know that there is nothing else like it anywhere on the planet.

The *Pioneer* topo map showed us that the whole Australia-sized Ishtar sits high above the global plains. To the west is the massive plateau Lakshmi Planum. The flat plains of Lakshmi cover more than seven hundred thousand square miles, an area twice as big as Earth's Tibetan plateau, at a height of more than two miles above the global

---

*Since *Terra* means "Earth," you may think this a funny name, equivalent to calling Earth's continents "Africa Venera" and "Asia Venera." We've given the different Venusian geographical forms names that reflect their classifications. The largest highlands are Terrae, intermediate-sized highlands are Regiones, low plains are Planitae, high plains are Plana, mountain ranges are Montes, canyons are Chasmata, and steep cliffs are Rupes.

average.* The perimeter of Lakshmi is composed of rough mountain ranges. The southern flank is a huge cliff, or "scarp." To the east, the Maxwell Montes jut out mightily toward the clouds. Rising to an astonishing height of thirty-eight thousand feet, Maxwell is by far the highest mountain range on Venus.[†]

Pioneer's low-resolution topo map raised many questions that only detailed imaging could answer. Is Lakshmi flat and smooth because of extensive lava flows? What kinds of mountains make up the ranges on its flanks, and how do they compare with terrestrial mountain ranges? Have the mountains been forced up by the motion of tectonic plates, perhaps advancing from an equatorial spreading center? Is the peak of Maxwell a gargantuan shield volcano? Does its large size imply that it is quite young, possibly still under formation, and has not had time to "relax" down under the pull of its own weight? On a more basic level, what caused the huge uplift of Ishtar Terra? Is it a continent in the terrestrial sense, a large raft of lighter rock floating on the denser mantle?

As *Magellan* peeled the clouds away from Ishtar noodle by noodle, the answers to at least some of these questions became clear. The surface of the Lakshmi plateau is fairly featureless even at *Magellan*'s resolution, but we can see that it is indeed made of huge volcanic flows. This plateau is broken only by a few volcanic calderas, the largest of which are my old friends Colette and Sacajawea, the two "craters" that, we can now see, clearly have a volcanic (not an impact) origin.

Very rough and dramatic terrain surrounds the entire plateau. All the large mountain ranges around Lakshmi are composed of long, curved ridges with narrow troughs in between. They resemble terrestrial mountain ranges, like the Appalachians, which were raised by compression of the planet's crust, and were probably formed in the same way. The Danu Montes, which form a seven-hundred-mile

---

*On Venus there is no sea level. Altitudes are often given as a height above the average global elevation.

[†]All of the surface features on Venus are named after female deities and historical figures, *except* the mightily protruding Maxwell Montes, which is named after a male physicist. That the only feature given this gender should also be the most phallic could provide a field day for Freudians and deconstructionist historians of science. But, as discussed earlier, this assignation has historical roots that predate the "women only" rule on Venus.

stretch of the southeastern perimeter, rise fifteen hundred feet above Lakshmi, before dropping off nine thousand feet to the plains. The ranges to the north and west have similar structures.

To the south a steep two-mile high cliff falls off from the Lakshmi plateau to the exterior plains. Its average slope is at least 20 degrees—very steep—and massive landslides seem to have occurred here.

To the east of Lakshmi, the ground begins rising up toward the great heights of the Maxwell Montes. *Magellan* showed that the south-western slope of Maxwell is astonishingly steep, gaining more than four miles of elevation over a horizontal distance of only six miles, an *average* slope over this huge distance of 35 degrees! There is nothing like this on Earth. The gentler eastern slopes fade into one of the largest areas of highly deformed tessera on the planet, Fortuna Tessera. This area extends to the east for twelve hundred miles, gradually falling in elevation until it fades under the surrounding plains.

Ishtar may be the best Venusian candidate for an Earth-style continent, an area of thick, low-density crust floating high above the surrounding plains. If it is a continent, it must be made of rocks fundamentally different from the basalts that dominate most of the planet. But there are difficulties here. One problem is that granite and other continental rocks may not form at all in the absence of water. Water is ubiquitous in the geochemistry of rocks on Earth, and we tend to take its presence for granted. But some geochemists have concluded that you need a certain amount of water to make the low-density continental rocks that float the continents of Earth. If this is the case, they may not exist on Venus, and our type of continents may be impossible. Also, the gravity maps of Ishtar, made by *Pioneer* and *Magellan*, indicate that it is not floating freely on low-density crust. Another problem is that on Earth continental rocks are made at subduction trenches where one plate slides beneath another and dives into the mantle. When the subducting plate partially melts, lighter, buoyant magmas rise to the surface, making the granite and other rocks that form continents. Some researchers believe that subduction cannot occur on Venus, because the high temperature makes the crust too buoyant. But, the fact is, the causes of subduction are not that well understood even on Earth. We scanned the new *Magellan* images carefully for evidence of subduction trenches. Some people think we found them, and some think we didn't.

If Ishtar is not a high-floating continent, then the tremendous height of Maxwell, dwarfing any mountain range on Earth, presents a great challenge. What could be holding it up? The lithosphere would not be strong enough to support such a weight unless it were hundreds of miles thick. But the lithosphere can't be nearly that thick if Venus has a heat flow similar to Earth's. For reasons discussed in Chapter 4, planetary size is related to thermal evolution, and thermal evolution controls lithospheric thickness. A lithosphere thick enough to hold up Maxwell creates more problems than it solves. Yet, there it stands, stubbornly ignoring geophysicists' pronouncements that it cannot exist.

It seems that Maxwell cannot be just sitting there, resting statically on a thick lithosphere or floating on low-density crust. Some kind of ongoing activity must be holding it up. The geological details of Ishtar suggest that it is currently undergoing severe compression, and whatever motions are crushing it together could be forcing up the highlands. This could be caused by a major area of downwelling in the mantle beneath Ishtar.*

Ishtar is often compared with Tibet, a massive region of compression, where India has been slamming into Asia for 50 million years and forcing up the Himalaya. The Himalaya probably reached its maximum height about 8 million years ago and is now starting to spread laterally, relaxing and flowing under its own weight. Why doesn't this happen to Maxwell? That's a good question. With its great height and incredibly steep sides to the southwest, you would expect such gravitational spreading before long. For this reason, we think that Maxwell (and perhaps much of Ishtar) must be quite young. The stiffness of the dry rocks of Venus might buy Maxwell some time, but some modelers have suggested that it might only be a few million years old—as young as the human race. The rugged topography of Ishtar may demand that like Earth, and unlike any other place in the inner solar system, Venus is currently an active planet.

It seems clear that crustal compression explains the main features of Ishtar, but it is not so clear that this compression is necessarily associ-

---

*This picture (compression due to downwelling) is similar to that which we invoked for the major areas of tessera, except that the downwelling would need to be much more vigorous and extend to deeper depths in the mantle to hold up a huge load like Maxwell. And whereas the tesserae are ancient, Maxwell may be quite young.

ated with downwelling in the Venusian mantle. Some have suggested that the best way to hold up Maxwell is the direct force of mantle *upwelling* beneath Ishtar. Also, the large volume of volcanism that flooded onto Lakshmi might have required a local hot spot. Once again we seem to have two opposite theories vying to explain the same structures.

It may not be as simple as a choice between upwelling or down-welling. Many areas of Ishtar show evidence of superimposed episodes of extension and compression. Maybe both upwelling and down-welling have shaped Ishtar, perhaps at the same time in adjacent locations, or perhaps at different times. Could this be telling us something about the kind of convection in the Venusian mantle? Maybe the patterns of flow vary from place to place over smaller distances and times than those on Earth, which tend to be stable over very large areas. We will encounter this idea again when we try to puzzle out the global pattern of tectonics on Venus.

MIGHTY APHRODITE

It took *Magellan* eighty days to map the 3,800-mile-wide Ishtar. It was now late November back on Earth, but people within the air-conditioned halls of JPL didn't necessarily notice that the air in Pasadena had become cooler and clearer. We were still on our first lap around Venus. At this point in the mapping cycle the attention of the *Magellan* team shifted three thousand miles to the south, where the western end of the other great Venusian highland area, Aphrodite Terra, began to appear. Africa-sized Aphrodite, the largest of all Venusian highlands, extends along the equator nearly halfway around the planet. It is shaped like a scorpion (see image in black and white insert and map in color insert) facing westward with its pincers threatening the lower rises of Eistla Regio and its mighty tail curving up through Atla Regio almost nine thousand miles to the east. You may recall that much of the original motivation for calling Ishtar a continent had come from the pre-*Magellan* scheme for Venusian tectonics, which involved a massive spreading center at Aphrodite Terra. This was thought to have caused plate motion to the north that resulted in compression and uplift at Ishtar.

Some researchers had looked at the cross section of Aphrodite on *Pioneer*'s map and seen the shapes of Earth's midocean ridges. Others had drawn very different conclusions. Some fairly specific predictions had been made as to what kind of highland Aphrodite is and the role it plays in the global tectonics of Venus, so we were eager to see what *Magellan* would reveal there.

It turned out to be a lot more complex than anyone had predicted. Aphrodite is not one continuous structure with a common history, but an amalgamation of many diverse uplifted areas, with three major regions of differing terrain. In this overall sense it resembles a terrestrial continent more than a midocean ridge (or spreading center). A continent—North America, for instance—may have regions of compression (the Appalachians), vast flat plains (the Midwest), regions of intense uplift (the Rocky Mountains), and areas of extensional faulting where the ground has been pulled apart (the basin and range area of Nevada). In contrast, a midocean ridge is basically one kind of structure with minor variations along its whole extent.

Western Aphrodite is composed of two giant uplifted areas of tesserae averaging more than two miles in elevation. Ovda Regio to the west is a plateau eighteen hundred miles across with steep sides that plunge abruptly into the surrounding plains. (See image in black and white insert.) Ovda has some of the roughest terrain on all of Venus. East of Ovda lies Thetis Regio, a slightly smaller tessera plateau, with somewhat less steep sides. Both these areas reveal, under Magellan's close scrutiny, a plethora of ridges, troughs, and domes amid the overall blocky, jumbled texture of tessera.

Both types of tessera formation models (upwelling and downwelling) have been invoked to explain the origin of Western Aphrodite. However, everyone agrees that at least this part of Aphrodite bears no resemblance to a terrestrial midocean ridge.

As we travel east from Thetis into Central Aphrodite, the topography changes markedly. Here we encounter a narrower highlands area, extending more than three thousand miles, where elevations rarely exceed one mile. This central region is split in many places by large, steep valleys. The deepest of these, Diana Chasma, with a depth of nearly two miles and length of six hundred miles, is on the same scale as the gargantuan Valles Marineris of Mars and much larger than any canyon on Earth. The valleys of Central Aphrodite often have high ridges on one rim and much lower topography on the other. Some have speculated that they are zones of subduction, where part of the lithosphere is being pulled below another. But here again an opposite theory is also in circulation. Some see an area of uplift and rifting instead, perhaps an incipient spreading center like Beta Regio or the East African Rift Zone.

As we travel still further east we enter the major uplift of Atla Regio, the tail of the scorpion, and the terrain changes again. Atla is a broad, elongated dome, nine hundred by fifteen hundred miles, averaging nearly two miles in elevation, with numerous volcanic features and abundant rifts. There are pancake domes, lava flows, and small shields. But the most impressive volcanoes here are two imposing shields, Ozza and Maat Mons. Maat Mons is the largest volcano on Venus and the second highest peak. At a height of more than twenty-five thousand feet and a diameter of one hundred eighty miles, it rivals Hawaii's Mauna Loa, Earth's biggest volcano.

Branching out from the center of Atla are several giant fracture systems, patterns of parallel, curving faults hundreds of miles wide and often thousands of miles long. Atla is uplifted in the center and the crust has been split in numerous directions, in a pattern resembling Beta Regio and probably indicating a major mantle hot spot. We have now traveled the length of Aphrodite and found no convincing evidence of the once-imagined spreading center.

East of Aphrodite lies an area of low plains broken by a nearly continuous band of huge rifts leading us from Atla back to the western slopes of Beta Regio. With a return to Beta, our circumnavigation of Venus is complete.

MOUNTAINS OF FOOL'S GOLD?

After *Magellan* had uncovered a large portion of the planet, we prepared and scrutinized global maps. Many interesting patterns that had not been evident from study of individual areas began to emerge from this global view. One of the most puzzling was this: the highest mountains of Venus are all surprisingly shiny. At altitudes above about thirteen thousand feet, the reflectivity jumps up and the ground abruptly gets very bright. Surface roughness cannot explain this, so something in or on the ground at these high elevations is different, making it highly reflective (see Figure 5.13).

5.13 Global Reflectivity Map: *This map, made with Magellan data, shows highly reflective areas as bright. A comparison with the topographic map in the color insert shows that the highest areas are also the most highly reflective. (NASA)*

Why do the high peaks shine so brightly? One property of the Venusian atmosphere may be important here. Because of the thick, sluggishly circulating atmosphere, the temperature is almost completely independent of everything but altitude. If you want to cool off on Venus, just head for higher ground. It hardly matters what time of day it is or whether you are near the equator or a pole. As a result, the altitude where the surface brightens corresponds to the same air temperature everywhere: 820 degrees Fahrenheit, or about 45 degrees colder than the surface. Maybe some temperature-dependent chemical reaction changes surface rocks to a more reflective mineral at 820 degrees. One candidate for the shiny stuff is pyrite, or iron sulfide (FeS_2). Some chemical calculations suggest that when the air cools to below this temperature, iron—which prefers other minerals at higher temperatures—jumps ship, grabbing sulfur out of atmospheric gases and forming pyrite crystals. Pyrite reflects radar much better than other iron-containing minerals. Even a thin coating on the high peaks of Venus could explain their increased reflectivity.

Pyrite is often called "fool's gold" for reasons that are obvious if you've ever seen a chunk of it. It has shimmery golden crystals that you can mistake for gold, especially if you are a hungry prospector on your last legs, searching desperately for that sparkling outcrop at the rainbow's end. But pyrite is a fairly common mineral on Earth, and perhaps on Venus as well. I have a lump of it sitting here on my desk. You have probably seen it for sale at airport gift shops, street fairs, and certain disreputable jewelery stores. You can buy it in any "rock shop" for next to nothing, a price that further validates the nickname.

Are the highest mountains of Venus fool's gold–plated? I like this idea. It's attractive for the same reason that the original equilibrium condensation theories of planetary formation were appealing. There is something almost irresistible about a simple and elegant theory that explains a large set of observations with some basic chemistry. But the pyrite theory has run into difficulties and sometimes acrimonious controversy. Some chemists have done experiments which suggest that pyrite cannot exist under Venusian surface conditions; it rapidly decomposes into other minerals. Others say these experiments are flawed because they don't properly account for the reactions between surface minerals and the sulfur-containing gases. The scientists on both sides are very gifted and accomplished. I don't know who is right.

There are other possible explanations. If this were Earth, and these were photographs instead of radar images, we would say that the high mountains are bright because they are covered with snow—that the altitude above which they all abruptly turn shiny is the snow line. Could we be seeing something analogous, a bizarre Venusian version of a snow line? Obviously, Venus is far too hot and dry for our kind of snow, which is made of frozen water, but maybe something else plays the same role there. On Mars, carbon dioxide condenses as frost (dry ice) in the winter. On Titan, there are clouds of liquid methane. Both these worlds are much colder than Earth, and things condense into liquids and solids that never would here. Is there something that is *always* condensed here but might be a vapor in the hot lower atmosphere of Venus and condense as "snow" on the cooler mountaintops? The substance would have to freeze at the right temperature and be highly reflective at radar wavelengths. Many elements that we regard as solids (because they *are* solid here on Earth) could be liquid or vapor in the hot depths there—things like tin, lead, copper, or zinc. Could one of these metals be frosting the peaks?

One element that may fit the bill is tellurium. This brittle, silvery metal makes up only one-billionth of a percent of our surface rocks, but we humans have found several industrial uses for it—for example, as an additive to improve the qualities of stainless steel. Tellurium condenses at a temperature and pressure coincident with the high-reflectivity line on Venus, and calculations have shown that a layer only one micron thick* would be sufficient to cause the observed brightness. If gases from Venusian volcanoes have the same tellurium content as those on Earth, there should be enough in the atmosphere to plate the high peaks.

One night recently I dreamt of a long road trip on an overcast day. I drove down a steep route that wound off a high plateau with a huge, strangely reflective mountain range in the distance to my right. When I awoke, I wondered if this scenic drive might be the road south from Ishtar Terra, down over the Danu Montes toward the plains below (although it seemed to be a mirror image, with east and west inverted). The massive mountains shining in the distance would be Maxwell. That made me wonder, would Maxwell and the other high peaks of Venus really look shiny to us, if we stood on, or drove over, the surface? We know they shine in radar images, but we don't know how they would look to our eyes, at "visible" wavelengths.

We have actual photographs of only a few spots, all below the altitude where the shiny stuff appears. If the peaks are naked surfaces of fool's gold, or tellurium plated, they would certainly sparkle, or at least shine back much of the diffuse light of the overcast Venusian day. But a thin coating of dirt or dust would dull this shine to our eyes without affecting the radar signal, which penetrates deeper than light. We will not know until we go and look, or at least send real cameras below the clouds to photograph large areas of the surface in visible light.

VENUS IN THE RAW

Why is it that chemical reactions causing micron-thin deposits could dominate the appearance of large areas of Venus but never could on Earth? There is one fairly stunning quality shared by all of the landscapes captured by *Magellan*. The whole surface appears to be remarkably well preserved, even though most of it is at least several hundred million years old. Virtually every geologic feature on the planet has the

---

*A micron is one-millionth of a meter, or 0.000039 inches.

appearance of being brand new, due to the near-total absence of the forces of wind, rain, or vegetation that begin to mask or destroy geological forms on Earth even while they are being created.

Everywhere on Earth you can see evidence of erosion. It is visible from orbital pictures or when standing on the ground almost anywhere. Rivers, sand dunes, rounded rocks, vast flat plains of sediments, and glacial valleys are just a few of Earth's ubiquitous erosional features. On Earth you can often judge the age of a mountain range simply by the state of erosion. The older mountains on the East Coast of North America are rounded and worn compared with the younger, more jagged Rockies. Here on Earth, erosion mutes and masks all but the youngest volcanic and tectonic features.

On Venus it looks like there is almost no erosion. There is probably no rainfall or vegetation, and the surface winds are sluggish. There are a few "rivers," as we have discussed; some patches of windblown dust; and even some sand dunes. But we see no evidence that erosion or deposition by wind has affected the shape of landforms, removed older structures, or covered wide areas with sediments.

In some ways, then, looking at Venus is like seeing an alternate Earth stripped of its layers of sediment, revealing the formation processes of rocks, mountains, valleys, and flows. This is the geological equivalent of one of those "visible man" (or woman) models with transparent plastic skin allowing you to see the internal structure of the body.

We stand to gain new insights from such an opportunity to observe geology in the raw, unsullied by erosion. It also creates new challenges. We need to adjust our methods and learn how to interpret what we see under very different planetary rules of operation. In each location a very long history is preserved, which can get confusing. Imagine taking a class where the blackboard is never erased between lectures. The information in each lecture might be very valuable, but after a while it may look like a mess. On Earth, erosion constantly erases the board, removing older structures. In some areas of the Venusian surface the evidence from many distinct tectonic episodes is "overprinted," or superimposed. This is potentially valuable, since we missed all the earlier lectures, but we have our work cut out for us in trying to read the board.

CHEMICAL WEATHER

Rust never sleeps.
—NEIL YOUNG

The slow erosion that does take place on Venus is probably dominated by *chemical weathering*. This also occurs on Earth, acting in concert with wind, rain, and ice, which make fast work of any rock formations that stand in their way. Chemical weathering slowly alters the properties of solid materials by chemical reactions with atmospheric gases. Rust is a form of chemical weathering, in which our iron creations are slowly reduced to dust by oxygen in the air.* Surface rocks are weakened by chemical weathering, leading to rockslides and collapsing of landforms. This takes much longer than most erosion we are used to.

If you've ever hiked or rafted in the Grand Canyon or anyplace similar, you may be impressed with the slow pace of erosion. River, rain, and wind have taken hundreds of millions of years to craft Earth's monuments. But chemical weathering on Venus is much slower yet, acting without mechanical scouring or the action of water as it seeps into the pores of rock, freezing, and thawing and dissolving minerals.

In the absence of a liquid solvent, like water, chemical weathering depends on diffusion, the random motions of molecules, to bring eroding agents into contact with the interior of rock. The rate of diffusion thus controls the timescale of this lazy process. You can imagine that it takes quite a while, even geologically speaking, for many molecules of atmospheric gas to invade the interior of solid rocks. Once inside, the gases react with the minerals, swapping atoms and gradually altering rocks chemical and mineral structure. Some of these reactions cause the rocks to expand or contract like the freezing and thawing of water on Earth. Even after the rocks are weakened in this way, they may still not have much incentive to crumble and fall if the surface of Venus is as calm and free of storms as we believe it to be.†

It may be the occasional Venusquake that does most of the moving of surface material; or perhaps ground-hugging currents of the

---

*Acid rain is another common form of chemical weathering on Earth, but like most terrestrial examples it is intimately mixed with physical erosion by the water that carries it, and we can't really separate the two.

†Of course, if there are occasional storms or strong surface breezes, we would probably not have detected them, because only a few landers briefly tested the winds before they fried.

thick hot atmosphere move downslope at certain times of day, taking loose material with them. The surface winds may be very slow, but it doesn't take a hurricane to move material when the air is ninety times as thick as ours. On a very long timescale, impact cratering also helps to shift loose surface material around. You would have to wait a long time at any given location, but in the absence of other processes, the winds and Venusquakes from large impacts, which probably move dirt around for a distance of hundreds of miles, will eventually affect every place on the planet.

Chemical weathering is so important on Venus, compared with the other terrestrial planets, for three reasons: (1) The other erosive processes of water (common on Earth) and wind (common on Earth and Mars) are nonexistent or slow, leaving only chemistry to do the work of turning rocks into dust. (2) Because of the high temperature at the Venusian surface, chemical reactions with atmospheric gases go much faster than they do here or on Mars, and gases diffuse into rocks faster because they have greater thermal speeds. (3) The atmosphere is chemically corrosive (more than on Mars, less than on Earth). As pointed out in Chapter 3, the sulfur-containing gases at the surface (which are probably continuously resupplied by volcanoes) are out of equilibrium with the minerals in the rocks, and this creates a lively environment for chemical reactions.

The near-absence of physical erosion and the dominance, by default, of chemical weathering greatly affects the way Venus looks in the *Magellan* images, and must also affect the way that Venus looks up close. In the eroded landscapes of Earth, we see the patterns of flowing water or blowing sand. Chemical weathering makes shapes that more closely trace the internal structures of rock. It finds and exploits planes of weakness that result from the crystal structure of the rock, or its history of faulting and folding.

No place on Earth is immune from physical weathering, but sometimes, when hiking in the Rocky Mountains, I come across spots where chemical weathering seems to dominate. Old rocks are just crumbling in place as though they've weakened until gravity has done them in. Sometimes you see a cliff slowly breaking up along cracks that seem to run deep in the fabric of the rock itself. A heap of the same type of rock lies beneath, often piled into what geologists call a *talus slope*. These slopes tend to be at a critical angle, the *angle of repose*, and will always return to it, crumbling spontaneously if they get any steeper. This is how I imagine that many places on Venus look. Chemical weathering weakens the rock, and gravity does the rest. Talus slopes are probably a common landform on Venus, but the small-scale textures that would confirm this are beyond *Magellan's* resolution. I hate to sound like a broken record, but this is another question that can be answered by surveying Venus with cameras.

SEA OF HOLES

After all the years of poring over the coarse Pioneer map and ground-based radar images, it's quite satisfying to look at *Magellan* images and clearly identify impact craters. Once we had mapped most of the surface, we could study the global distribution of craters and look for clues to planetary history.

In rural areas of the United States where guns are commonplace and youth are bored, you can roughly date the road signs by the number of bullet holes found in them. The older signs are more shot up (see Figure 5.14). It's the same with craters and planetary surfaces. Impact craters are found nearly everywhere in the solar system; they are the universal chronometers of planetary science. In the absence of other processes, a planet will gradually accumulate craters from the random impacts of asteroids and comets. We know fairly well how many of these stray bodies exist and how often they hit planets, so crater density can be used to calculate the age of a surface. In *Magellan* images of Venus we have counted just over nine hundred impact craters. This corresponds to an average surface age of half a billion years, give or take a couple of hundred million. This makes it by far the youngest planetary surface in the inner solar system, except for Earth's.

Comparisons of crater density in different areas and different types of terrain help us assemble a historical sequence for a planet.* Older, less active areas will have collected more craters. They have been wiped out on younger, more active surfaces. For example, the relatively young northern plains of Mars are much more sparsely cratered than the ancient southern highlands. This technique works great on most

---

*By crater density of an area, I mean the number of craters per square mile.

5.14 A Heavily Cratered Surface: *Planetary surfaces accumulate craters with age. In some places road signs accumulate bullet holes. (Tory Read)*

planets. It has helped us learn the history of Mars and the Moon. But it doesn't work on Venus.

Venus is "impact challenged." Its surfaces lack small craters. It's easy to see why if you imagine you are standing in a few inches of water and throwing rocks of different sizes down at the sand below the water. With pebbles, you cannot make a dent in the sand below because the water slows them down and they fall gently to the bottom. A large boulder that you can barely lift will make a nice little crater because the water cannot stop it. Now if you change the level of the water, you change the critical size between rocks that can and cannot make craters. The more water, the bigger a rock you need. Similarly, the thicker a planet's atmosphere, the larger an impacting object has to be to survive passage through the air at high velocity and make a crater. Every planet with an atmosphere has a cutoff size, depending on atmospheric thickness, below which there are no craters on the surface.

On Venus, the smaller impacting objects that would otherwise do most of the cratering are filtered out by the thick atmosphere. They burn up or explode before they hit the ground, just like meteors do on Earth. On Earth, only dust and pebbles get snuffed (making "shooting stars"), but on Venus you don't get any craters smaller than about two miles across. We expected this. In fact, the small size cutoff of craters conforms so closely to pre-*Magellan* predictions that it was a confidence booster for our models of the passage of small bodies through planetary atmospheres. These models were fine-tuned after July 1994, when nature kindly arranged an unprecedented opportunity to observe several large impacts as fragments of comet Shoemaker-Levy 9 plunged into Jupiter. Scientists used observations of these events to improve their models of high-speed atmospheric entry on Venus, which allowed the crater counters to refine their estimates of the surface age.

Any well-cratered planetary surface has many more small craters than big ones, because there are many more small stray objects in space than big ones. Although these holes are rather small, we have to count them all. The overall peppering of Mars and the Moon with very small craters is what allows us to determine the relative ages of different areas. But since Venus has no small craters, it doesn't have enough of them to help us with chronology. All areas of the surface have more or less the same low number of craters. There are some minor variations in crater density from place to place, but these are of the kind that we would expect to be produced by the blind marksman of impact cratering. The global map of craters on Venus (see Figure 5.15) is indistinguishable from a random pattern.

It turns out to be pretty tough to reconstruct a history for Venus, at least using images taken from orbit. Many of our usual techniques just don't work. Crater counting is out, and erosion is pretty much useless, since basically nothing is eroded. (On Mars older features are noticeably more heavily sandblasted by wind and buried by dust than younger ones.) So we must rely heavily on other means to assemble a chronology of the surface. In some places it is obvious, on a local scale, that a particular flow or episode of faulting came earlier or later, but efforts to construct a global history have been controversial. For example, it is clear in individual locales that the tesserae predate the smooth plains, which lap up onto them. But does this mean that *all* tesserae are older than *all* plains, everywhere around the planet? Did tessera formation stop everywhere before plains formation began anywhere? On any other planet we could just count craters and find the answer. On Venus, we must coax the answers from more subtle clues.

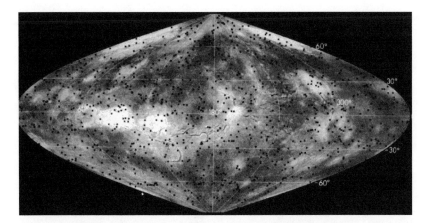

5.15 Global Distribution of Craters: *This map has a dot superimposed for the location of every impact crater identified in* Magellan *images. (R. Strom)*

There is another reason why the history of the Venusian surface has been tricky to decipher. It turns out to be a most unusual one, with a surprising distribution of surface ages compared with any other planet. Our first clues to this bizarre history came when we noticed that something is not quite right about the impact craters.

WHAT IS EATING THE CRATERS OF VENUS?

All the craters on Venus look unnaturally pristine. Instead of blending into the volcanic plains that cover most of the planet, they seem planted on top as an afterthought, as though a crew had built a cheap movie-set planet and realized at the last minute that they had better throw in some craters (see Figure 5.16).

The total number of craters indicates that the planet's surface is not very old by planetary standards—a billion years, tops. Therefore, some process has removed most of the craters that ever formed on the planet. But there are *no signs* of any such process. The thief who stole Venus' craters has successfully covered his tracks. There are three suspects, but each has a good defense.

If it were *erosion*, we would see craters in a range of states, from pristine fresh ones to highly degraded ones that are nearly gone. We see that on Mars, where erosion by windblown dust is a dominant process.

5.16 The "Crater Farm": *Three fresh-looking impact craters lie atop older fractured volcanic plains. The largest of these is forty miles across.*

Craters can also be destroyed by *tectonic disruption*, if they are simply cracked and faulted to the point where they are no longer recognizable. Here again, we would expect to see the process at work, but the number of partially disrupted craters on Venus is small.

A prime suspect is *volcanism*, since volcanic flows dominate the visible surface. But if craters were mostly being covered over by lava, then we should see a lot of them partially buried beneath the plains. For example, the volcanic plains of the Moon contain craters ranging from those with barely detectable circular outlines that have been almost completely covered by lava to those that are almost pristine but have lava flows lapping slightly up their flanks. On Venus only about 4 percent of the craters are partially covered by volcanic flows, and almost none are mostly buried (see Figure 5.17). This seems to rule out volcanism as a dominant process removing craters from Venus. Why, on a planet smothered with volcanic features, are the craters so untouched? Do volcanoes on Venus worship craters, as Hindus vener-

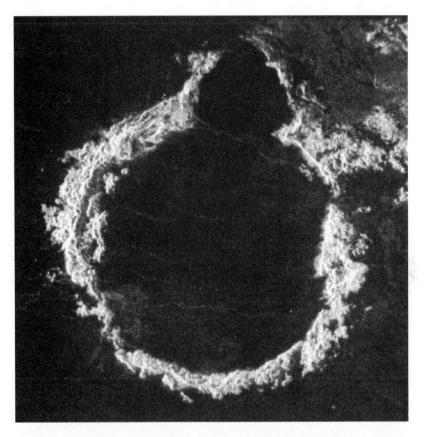

5.17 **Partially Buried Crater:** *This twenty-five-mile-diameter crater has been largely covered by later lava flows, so only the rim remains. Such features are surprisingly rare on Venus. Most craters postdate the flooding of the volcanic plains.*

ate cows, and thus spare them from their otherwise global carnage? This is not a favored hypothesis, but we needed a way to explain the odd, and globally pervasive, occurrence of pristine craters overlaying widespread lava plains.

It is as if some mysterious process were swallowing craters whole and leaving no trace. There are a few ways in which we can imagine this appearance coming about. If you see a road sign full of holes, it is not necessarily old. Some well-armed fool pumped up on testosterone and booze could have opened fire on it the night before last. What if all of the craters really were added as a kind of afterthought, produced by a very recent barrage of impacts? On any planet with reasonable erosion rates, you could test this idea by noticing if some of the craters were worn down, showing signs of age. (Are the bullet holes rusted?) But on Venus you can't tell, so the hypothesis is consistent with the way all the craters sit atop the "paint job" of planetwide lava flows.

Unfortunately, this scenario is only slightly more plausible than crater-worshiping volcanoes. Everything we know about the inner solar system suggests that there has been no recent burst of cratering activity. Impacting objects do not discriminate greatly among the terrestrial planets. If there had been a recent shower of large meteoroids, we would also see its effects on the surfaces of the Moon, Mars, and Earth. We don't.

One scientist suggested that Venus could have had a Moon that disintegrated relatively recently, showering the planet with fragments that made most of the craters. This theory is hard to disprove, but unsatisfying. We don't like to resort to explanations that require extraordinary or fortuitously timed events unless we have exhausted all other possibilities. You can always invent a contrived theory to explain any observation, but the universe in general does not seem designed like a Rube Goldberg machine, and if we dig deeper we can often find simplicity. This search for simplicity is often called Ockham's razor.*

VENUS GETS A COMPLETE MAKEOVER

Here is another way to explain the unusual crater population: Suppose that half a billion years ago something suddenly happened to Venus, wiping out all older craters. Vast lava flows occurring simultaneously all over the planet would do the trick. Then, if there has been relatively little surface activity since that time and Venus has been slowly collecting craters all along, things should look as they do.

This idea, called "catastrophic resurfacing," rubs many scientists the wrong way. Like the disrupted moon theory, it invokes a special event, and quite an incredible one at that. We can hear Ockham sharpening his razor. Some scientists would prefer a "steady-state" model—

---

*This principle is named after William of Ockham, who said, in the fourteenth century, "It is vain to do with more what can be done with fewer." We take this to mean: "Why assume that things are complex if a simple theory can explain all of the observations?" We scientists are guided by a deeply held conviction that there exist very simple laws, which we can discover, governing the behavior of much of the universe.

one that employs ongoing processes operating at more or less constant rates and that does not require us to accept that the planet changed fundamentally in the geologically recent past. Thus many vigorous arguments have been offered in valiant attempts to save us from having to accept catastrophe. Perhaps craters have been steadily destroyed all along by volcanic and tectonic disruptions that are just the right size to destroy whole craters but not partial ones. If small patches of the surface are continually destroyed in this way, eventually the whole surface will be affected and no traces of older craters will be left.

We have developed new computer models that simulate the evolution of the surface of Venus to test these competing ideas. We call these "Monte Carlo" models because they employ random number generators like digital dice to simulate random processes in nature. The results have confirmed what seems intuitively correct: if craters are forming at random and are being destroyed by volcanism (which is also randomly occurring over a long period of time), then we should see a lot more craters partially covered by lava flows. To return to the bullet-holes-in-road-sign analogy, if someone is going around repainting signs that have been shot up, then holes that have been painted on should be common. If not, the painting must have stopped before most of the shooting. Monte Carlo models seem to support the catastrophic resurfacing theory.

The initially disturbing idea of catastrophic resurfacing became more palatable when it received support from other lines of reasoning. Models of the interior evolution of Venus suggest an oscillation between periods of relative quiescence and periods of instability and rapid surface overturn. In other words, Venus may "freak out" occasionally, getting rid of its internal heat in great planetwide spasms of activity, rather than the steadier cycling of lithospheric plates we are used to here on Earth. These models predict intermittent bursts of greatly enhanced volcanic activity at intervals of several hundred million years. Is the observed crater population simply an accumulation of impacts since the last time Venus turned inside out? The researchers who made these models did not set out to prove that catastrophic resurfacing has occurred. In fact, some of them were quite skeptical at first. They just set up some equations based on the laws of physics and chemistry, made educated guesses about the interior structure, pressed the start button, and watched Venus oscillate between long periods of calm and brief spells of rapid surface activity. A wild idea like catastrophic resurfacing is less likely to be shredded by Ockham's razor if it turns out to be an inevitable result of physical and chemical evolution.

The initial proposal of the catastrophic resurfacing model drew strong reactions, for and against, from many planetary scientists. Several years of frustrating debates had the two sides repeating the same arguments, and it wasn't clear that anyone was listening. Perhaps the community became briefly stuck in this cul de sac because of the power of the loaded word *catastrophic*. It recalls the conflict between catastrophic and uniformitarian views of Earth history, described in Chapter 4. Geologists have spent much of the last several centuries demonstrating that we can explain surface features on Earth without recourse to a biblical flood. Must we now accept that much of Venus's surface *was* created in one great volcanic flood of similar proportions? The search for a steady-state model to explain the craters of Venus can be seen as an attempt to find a uniformitarian alternative. Perhaps the choice of the phrase "catastrophic resurfacing" was unfortunate, and "sudden global resurfacing" would have provoked a more constructive debate (although it does not sound as scientific).

The debate also took on overtones of the cultural conflict within planetary science mentioned earlier. Early advocates of catastrophic resurfacing tended to be descriptive geologists, while their "opponents" often came from the more quantitative field of geophysics. The geologists had good arguments based on the overall appearance of the surface, but their quantitative statistical analysis of crater distribution was in some cases lacking. Their opponents sometimes attacked the statistics while ignoring the basic arguments.

The repetitive debate seems to have largely died down, to the great relief of most Venus scientists. Although some controversy continues, the community seems to be converging toward an acceptance of something like catastrophic resurfacing (although not everyone calls it that). Roughly 600 million years ago, Venus seems to have been wiped clean of craters.* Some combination of widespread volcanic flooding and tectonic disruption created a *tabula rasa* for later cratering. Most

---

*The crater counts give a surface age of between 300 and 800 million years, with a best guess of 600 million years. I don't want to keep saying this, so I will just refer to the age as 600 million, or sometimes half a billion years. What's a hundred million years among friends?

of this activity died down quickly, producing a planetary surface that may be nearly all the same age. But planets are never this simple. Even if Venus did receive a near-complete makeover 600 million years ago, any complete picture must include the greater complexity inevitable in an Earth-sized planet.

A lower rate of volcanism and continued tectonic activity has continued in certain areas to the present day, producing the small number of flooded and disrupted craters. Several lines of evidence for this include fresh flows on some volcanoes, particularly the large shields (although the lack of erosion *could* fool us into thinking older flows are fresh) and the nonequilibrium mix of atmospheric gases, with an excess of SO_2 probably supplied by active volcanoes.

Newer, more detailed analysis of the global crater distribution also points toward more recent surface activity. Although no large areas of lower or higher crater density can be found, some *types* of terrain have more craters than others. One way to think of this is to imagine that you could rearrange the surface so that all the shield volcanoes (say) were next to each other in one large continuous area, rather than spread around the planet. When we do this, we find that large volcanoes have fewer craters than other areas: they seem to be, on average, only half the age of the global plains. We have tried this trick with some other kinds of terrains with mixed results. The highly deformed tesserae are apparently the oldest areas of Venus. They may be the only remnants of the surface that existed *before* catastrophic resurfacing—high-altitude "islands" that escaped the global volcanic flooding. If so, then perhaps they should have more craters than other areas. This is hard to prove, possibly because craters are hard to pick out on images of these highly cracked and wrinkled surfaces. Compared with the smooth plains, searching for craters on the tesserae is like playing "Where's Waldo on Venus?" Many distracting details make it hard to find what you are looking for.

Venus seems to have changed not only its rate of volcanism but its style, fairly abruptly, around 600 million years ago. Before then the planet was vigorously repaving itself with vast plains of basalt. Then, for reasons we don't understand but that must be related to the evolving interior of the planet, it switched over to a lower rate of volcanism, concentrated in specific areas where hot spots forced up broad domes in the lithosphere, creating giant shield volcanoes.

If this general picture holds up, the implications for the evolution of the planet are staggering. To cover up all preexisting craters, an episode of rapid volcanism would have had to flood most of the planet's surface with lava to a depth of three to six miles. This implies a rate of volcanism fifty to one thousand times higher than the recent rate, depending on how long the resurfacing lasted.

Once again we are forced to accept that some aspects of planetary evolution have been anything but uniformitarian. Our recently increased appreciation for the role of very large impacts in planetary origins and evolution should make this easier to swallow. But large impacts are outside invaders; whatever wiped out the surface of Venus came from within.

Although the idea that our planetary twin had such a radical makeover as recently as 600 million years ago may be unsettling, it would be instructive to remember that at roughly the same time Earth was undergoing its own midlife crisis. For, back on Earth, this was the age of the Cambrian explosion, which dramatically transformed our planet from a world on which nothing more complex than bacteria and algae had evolved for billions of years to one teeming with an incredible diversity of animal and plant species.

Perhaps the tesserae, the ancient, highly fractured areas that make up 10 percent of the surface of Venus, are analogous to the ancient Precambrian areas of Earth, such as the Canadian Shield, that predate this metamorphosis. Earth's Precambrian areas have certainly been through a lot of stress and strain over hundreds of millions of years. If they were not eroded and covered by sediments, they might look much like the tesserae of Venus.

DURING THE FLOOD: VENUS IN AN ALTERED STATE

It is hard to imagine what Venus might have looked like during the great flood, had there been anyone to look. Much of the present volcanically dominated surface may have been in active formation at that time, with giant flows of molten basalt occurring simultaneously around the whole planet.

This furious surface activity must have had enormous effects on the atmosphere and climate. We haven't worked out the details of this, and it won't be an easy task, although it should be fun. But I would wager my left brain on this statement: you do not completely resurface a planet like Venus without causing profound global change. No one really knows what kind of change, so don't believe anything anyone says about this. However, the first attempts to model this have produced some pretty colorful results.

Venus today has a greenhouse effect gone wild. We think that volcanoes are actively supplying some of the trace gases that maintain this massive atmospheric greenhouse, especially water vapor and sulfur dioxide. (The same gases are commonly found emanating from volcanoes on Earth.) If volcanism was hundreds of times more active for a spell 600 million years ago, the atmosphere would have been flooded with hundreds or thousands of times as much of these greenhouse gases.

Especially because of the abundance of water vapor, the surface would have been much hotter than it is now (hard as this is to imagine)—how much hotter, it's hard to say. A new generation of improved models will help to answer these questions. In any case, the catastrophic resurfacing hypothesis implies that half a billion years ago Venus was a very different world. That's not a long time in a solar system that is 4.5 billion years old. It is as if a forty-five-year-old person had taken on a whole new personality five or six years ago.

One area of study that we need to reconsider is the evolution of water on Venus. In Chapter 3 I discussed the debates about whether the meager amount of water currently found on Venus (thirty parts per million in the atmosphere) is a remnant of an ancient, Earthlike ocean that escaped into space or the result of more recent injections into the atmosphere by icy comet impacts or volcanic eruptions. Now it seems that this argument may be moot. If planetwide volcanism flooded the atmosphere with water 600 million years ago, the small amount we observe today could stem mostly from this major global hiccup and have little to do with either ancient oceans or recent comet impacts.

IS VENUS DEAD, ASLEEP, OR JUST CONFUSED?

Now that we have heard and seen some of the evidence, let's return to one of the central questions that has framed Venus-Earth comparisons: Does Venus have a system of global plate tectonics like Earth's? If not, why not, and how does she rid herself of internal heat? You can look at this as an accounting problem: Earth's internal heat engine loses energy at the rate of about 30 trillion (3×10^{13}) watts. That's a lot of light bulbs. This energy comes mostly from radioactive decay, with a smaller (and somewhat uncertain) contribution from the deep lingering heat of birth 4.5 billion years ago. It all comes out as heat flow at the surface. Plate tectonics is responsible for most of this cooling. Heat rises to the surface with magma at midocean ridges. About 70 percent of Earth's heat flow is due to subduction: slabs of cold lithosphere plunging into the mantle like ice cubes into a hot drink. But other mechanisms are also necessary. Rising mantle plumes of hot material, which make hot spots like Hawaii, probably take up much of the slack.

Venus should have a similar overall heat production, and it has to be losing this heat somehow. It can't be holding it inside because the whole planet would just be molten. It is not enough to simply pronounce that Venus looks different from Earth, so it doesn't have plate tectonics. We have to find some cooling mechanism.

Magellan images certainly show abundant evidence of tectonics. Many regions show evidence of large-scale patterns of stress that have pulled apart or squeezed together areas of the planet's crust, creating mountain chains, extensive patterns of ridges out on the plains, and rift valleys. But the global pattern does not closely resemble terrestrial plate tectonics. We don't see abundant subduction zones or a globally interconnected system of spreading ridges.

Also, the age distribution of the Venusian plains seems to be quite different from that of the seafloor on Earth. To a first approximation, the plains of Venus seem to almost all have the same age, about 600 million years. On Earth, there is a clear pattern of increasing age as you travel away from a midocean ridge, and all of the seafloor is younger than 200 million years. We do have to keep in mind that we don't know exactly how plate tectonics might look on a planet with no ocean and a 900-degree surface temperature.

I will now summarize five different ideas that have been proposed to make sense of the global tectonics of Venus. They are not all mutually exclusive, and a combination of some of these may eventually provide the answer to how Venus stays cool.

1. PAPER PLATE TECTONICS

We find clues in the way that various types of landforms are spread across the face of the planet. The distribution of volcanoes seems important. Although there are some zones of concentrated volcanic activity, the various volcanic forms are widely dispersed, spread out over large areas of the planet. On Earth, by contrast, volcanoes are concentrated along the boundaries between tectonic plates (like the Cascades or the Andes).

Tectonic features are also more dispersed on Venus. On Earth, the plate boundaries are where the action is: much of the interesting surface relief and almost all of the volcanic and seismic activity are confined to the narrow belts where plates interact. On Venus these activities seem to be distributed over larger areas. The forces pulling at the crust on Venus have apparently stretched large areas in various ways instead of forming clean breaks at plate margins.

This difference may be related to the surface temperature. As we have discussed, all solid materials become less viscous, or more *ductile* (squishy), when heated. On Earth, the "brittle-ductile transition," the depth where it is hot enough for rocks under stress to deform rather than break, occurs so deep that the lithosphere above it moves in large solid plates, allowing plate tectonics as we know it to occur. Perhaps on Venus, where the surface rocks are already halfway to their melting points, this ductile behavior begins much closer to the surface. Then the lithosphere would be thinner and more pliable. Forces resulting from convective motions in the mantle might tend to stretch and compress areas of the crust rather than float them around and crash them together. The wide zones of tectonism and volcanism we see may represent more spread-out, distributed plate boundaries, in contrast to the narrow, linear zones of Earth.

So Venus may have a kind of "distributed plate tectonics," with a larger number of smaller, less rigid, and less well defined plates that deform more easily than those of Earth. The plates of Venus are not like solid barges that crumple and break only where they come together at the edges. They may be more like "paper plates" that just stretch and buckle all the way through when they collide or are pushed up by a rising mantle plume. On Earth, plate tectonics results, over time, in large, coherent horizontal movements over thousands of miles. On Venus, the smaller plates may just jostle for position, staying more or less put and deforming in place rather than moving large distances.

This idea places Earth's global tectonics somewhere in between those of Venus and Mars. Mars is described as a "one plate planet," without much geological activity, where the lithosphere is just too cold and thick to break up into plates. Earth's lithosphere is thinner than Mars's and its interior more active, so it has about a dozen plates that move around, with well-known results. Could Venus just be a bit farther in this same direction, with an even thinner lithosphere broken into a larger number of "paper plates" that do not act rigidly?

When it comes to global tectonics, the three planets may present yet another Goldilocks problem. If I remember the story correctly, after sponging a meal off of the Three Bears, Goldilocks headed upstairs for a snooze. Maybe the three beds she found there are analogous to the lithospheres of the terrestrial planets: Mars is *too hard*, Venus is *too soft*, and Earth is *just right* for our kind of plate tectonics.

2. INTERMITTENT PLATE TECTONICS

There is another—more radical and in some ways more exciting—idea about the long term evolution of Venus. Perhaps Venus has intermittent plate tectonics, a more Earthlike system but one that periodically switches on and off.

Another puzzle of the Venusian surface lends some support to this idea. Although much of the planet is composed of flat volcanic plains, a few regions rise steeply to great heights. As I've discussed, it's damn hard to think of a mechanism to hold up the thirty-eight-thousand-foot Maxwell Montes and the other high peaks. If the lithosphere were thick enough to hold them up, it would be insulating the interior of the planet like a thick pile of blankets. But then heat would build up inside. The interior could heat up only so much until the lithosphere would eventually thin.

This might be precisely what is happening now. When the lithosphere gets thin enough, another episode of plate tectonics could begin, accompanied by vigorous tectonic and volcanic activity that would release heat, cool the interior, and start the process over.

In this model the internal heat flow of Venus operates like a thermostat-controlled building heated with occasional blasts from a hot furnace, rather than by a steady stream of warm air. Venus may not have a relatively smooth and steady heat flow and tectonic system, like Earth. Instead, it may release its heat in great spasms of vigorous surface activity. The most recent period of frantic tectonic and volcanic activity, the last furnace blast, may have been 600 million years ago. This is one of the "oscillating interior" models of Venus mentioned earlier. It is conceivable that during the period of catastrophic resurfacing, Venus had a system of plate tectonics resembling Earth's. The low level of volcanism we see today, manifested mostly in hot spots, may represent the way Venus behaves in the intervals between these great spasms. The correct answer to the vexing question "Does Venus exhibit plate tectonics?" might be: "Yes it does, occasionally!" If we want to see plate tectonics on Venus, we may just have to wait a few hundred million years.

Why should Venus oscillate this way if Earth doesn't? We are still trying to understand the way that plate tectonics works on Earth. Often scientists begin their talks about Venus with ideas and speculations about how plate tectonics *might* work on Earth and then proceed to discuss possible differences with Venus. The different behavior might have to do with our planet's great abundance of water and the role it plays in "lubricating" plate tectonics. Earth's crust and mantle are hydrated: a lot of water is bound up in the structure of the rocks. At the base of Earth's lithosphere, hydrated rocks contribute to a partially melted zone with low viscosity. This thin zone may allow plate tectonics on Earth to be self-regulating. If heat starts to build up a bit, the viscosity of this layer falls (it flows more easily). As a result, convection speeds up, increasing heat flow and cooling the mantle slightly. This feedback mechanism, a kind of thermostat, may be what keeps Earth running smoothly.

Some models suggest that without this hydrated layer, plate tectonics might not be self-regulating. In this case, Earth's "engine" of plate tectonics and mantle convection would run less smoothly. Heat could build up over long periods of time until it forced its way out in great convulsions. In other words, without a lot of water in the crust and mantle, Earth would behave more like Venus. If Venus, because of her dry condition, has trouble letting off steam gradually, she may throw occasional tantrums instead.

3. IS VENUS PAST ITS PRIME?

Is the planet now sleeping between episodes of widespread global tectonics? Or did Venus just up and die? Maybe it once had continuously hyperactive interior convection and vigorous surface activity, but then shut down permanently 600 million years ago because it had cooled so much.

This idea runs into difficulties because the low abundance of radiogenic argon-40 in the atmosphere (discussed in Chapter 3) sets limits on the possible total volcanic activity over the planet's history. The argon level is incompatible with a continuously high level of volcanism until comparatively recently. It is more consistent with a brief spasm.

Plus it would be a bit odd for vigorous volcanism to have persisted so close to the present day and then stopped permanently.

4. BURIED PLATE TECTONICS

Here is an idea weird enough to just possibly be true. Maybe Venus has a vigorous system of lithosphere recycling, perhaps even one resembling terrestrial plate tectonics, but it is all going on beneath the surface. A surface layer that is buoyant and stiff, because it is hot and dry, could be sitting on top of plates that move around below, pushing and stretching it into the patterns we see.

This would help us out of the following quandary: There is plenty of evidence for new crust being created on Venus through volcanic outflows and rifting. But where is the old crust going? There may be little evidence for subduction. If the surface layer is everywhere too buoyant to be subducted, maybe subduction is occurring only at the bottom of the lithosphere, with large pieces of plates peeling off from below and rejoining the mantle. This process, called *delamination,* also occurs on Earth, and some scientists believe it is very important in Earth's overall heat flow, second only to subduction. If delamination is widespread on Venus, it could cool the interior without leaving obvious marks, like subduction trenches, on the surface.*

5. DISORGANIZED PLATE TECTONICS

On the other hand (you may be wondering by now just how many hands we need to hold all of our ideas about Venusian tectonics), some scientists *do* see evidence of subduction at certain locations in *Magellan* images. There are many arc-shaped trenches that are asymmetrical, with an uplifted rim on only one side, like subduction trenches on Earth's ocean floors. One of these is Artemis Chasma, the huge trench that arcs around the southeast of giant Artemis Corona in south-central Aphrodite (see Figure 5.18). Many large coronae are

*My wife says this should be called "lithosuction."

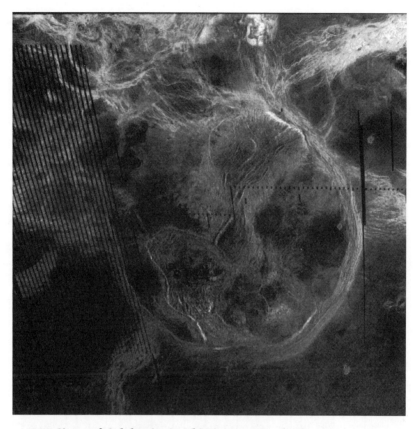

5.18 Signs of Subduction?: *This is Artemis, the largest corona on Venus at sixteen hundred miles across. The trench surrounding it (called Artemis Chasma) could be a subduction trench. (The black lines are missing "noodles" in* Magellan's *data.)*

partly surrounded by trenches that could be local sites of subduction. It may be that after an uplift of hot rock forms a corona, the cooling material subsides around the margins in a local subduction event.

These possible subduction trenches are often near areas of obvious mantle upwelling, such as coronae or rifts, suggesting that on Venus areas of upwelling and downwelling may often occur in close proximity. This would be quite unlike Earth, where subduction zones occur far from the uplift, rifting, and spreading at midocean ridges. But horizontal plate motion on Venus, if it occurs at all, is probably more localized. If subduction is difficult because of a buoyant or rigid crust, then rifting caused by hot spots might first be necessary to "break the ice," leaving cracks in the lithosphere where subduction can occur later.

This view is consistent with the "paper plate" model described above and with a pattern of mantle convection on Venus that is divided into smaller areas of upwelling and downwelling, less organized on a large scale than on Earth.

Although it is clear that an organized system of Earth-style, large-scale plate tectonics is not currently operating on Venus, many of the processes we associate with plate tectonics may be occurring on a more local scale. This could reflect a difference in lithospheric thickness or interior convection style between the two planets or, most likely, both.

As you can see, we are still searching for the overall pattern of global tectonics on Venus. At this point the ideas themselves are like tectonic plates floating on a deep mantle of mystery, cracking and flooding at the edges where they jostle against one another. Some will be subducted and disappear, some will remain, perhaps melding together as the margins between heal over and a larger, more complete theory emerges. The jury is still out, but at least now our deliberations are more informed. We've seen the evidence even if we don't know for certain what it all means. Our studies of Venus may be in a similar state to earth science in the 1950s before the wide acceptance of plate tectonics. We understood the regional geology of many areas but still lacked the global picture.

One basic difference among the competing theories is that some depend on Venus having a thin lithosphere and some require a thick one. So which is it? Is Venus thin-skinned and losing heat rapidly, or is she resting between global convulsions and gathering internal heat under a temporarily thick skin that is holding up the mountains? One way to probe the structure of the lithosphere is with gravity mapping, which tells us the distribution of mass below the ground.

By late September 1992, *Magellan* had completed three mapping cycles and imaged 98 percent of the planet. The plan was to use the rest of the spacecraft's life to map gravity.* *Magellan*'s initial highly elliptical orbit, optimized for radar imaging, was not good for gravity mapping. That requires a low circular orbit, which brings the spacecraft closer to the whole surface and the pull of the interior bumps of density beneath. We had a well-developed plan for some bold and innovative

---

*We already had a crude gravity map from *Pioneer*, but with *Magellan* we hoped to do better.

spacecraft maneuvering to put *Magellan* on such a circular orbit. But first it simply lowered its orbit slightly and made one more pass around Venus (cycle four) to gather some gravity data. Since the new moves were risky, we wanted to be sure to have at least some kind of gravity map if the spacecraft was lost.

LIVING DANGEROUSLY

The cycle 4 gravity survey was complete by late May 1993. *Magellan* had attained most of its goals. There was less to lose now, and the *Magellan* team was ready to alter the spacecraft's orbit for gravity mapping. They had to fight like hell to keep NASA from cutting off funds for this. NASA was simultaneously battling Congress for funding to keep other missions alive, but killing *Magellan* at that point would have been sad. It's hard enough to get a spacecraft flown to another planet. Why shut it off while it is still there gathering priceless data, just to save a small fraction of the total cost? It was touch and go for a while, but, operating with a reduced staff and no margin for error, the last phase of *Magellan's* journey began. It involved some tricky spacecraft maneuvers never tried before, the spacecraft engineer's equivalent of extreme sports.

Magellan did not have enough fuel to alter its orbit in the desired way simply by firing its thrusters. Instead, like a skateboarder dragging a foot to brake herself, the spacecraft dipped repeatedly into the upper reaches of the Venusian atmosphere, dragging against the thin air to slow itself down and achieve a new orbit. This technique, known as aerobraking, had been on the books (including a lot of science fiction)* for a long time, but *Magellan* was the first to try it.

We had to have a good idea of the structure of the upper atmosphere so that we could calculate the force it would exert on the spacecraft. Fortunately, we had been able to observe the final few orbits of *Pioneer* in 1992 as it helplessly dipped into the atmosphere before losing control and plunging to its death. Measurements of the atmospheric drag on *Pioneer* helped us plan *Magellan's* aerobraking moves.

The controllers at JPL fired *Magellan's* thrusters, pushing the low part of its orbit down to a height of eighty-six miles, within the upper

---

*In a dramatic scene of the movie *2010*, the crew endures a harrowing aerobraking maneuver through Jupiter's atmosphere.

reaches of the atmosphere. They also changed the orientation of the craft to maximize air resistance by letting the big main antenna trail behind and setting the big solar panels flat on to the breeze. I'm sure Ferdinand Magellan would be proud to have his name associated with this daring new navigational trick that allowed us to sail through the uncharted waters of the outermost Venusian atmosphere.

During repeated passes from May to August '93, atmospheric friction robbed the spacecraft of orbital energy, pulling down the high part of the orbit. Then thrusters were fired to raise the low point out of the atmosphere. Aerobraking worked. We achieved a lower, more circular orbit, ranging from 110 to 340 miles above the surface. From these heights, the motions of the spacecraft responded more sensitively to interior variations in density. From August '93 until its death in October '94, *Magellan* orbited Venus more than six thousand times, feeling out the subtle changes in gravity from place to place around the planet. Recording each wiggle of the spacecraft's motion, we produced an improved gravity map of the whole planet.

The gravity observations have not yet resolved the question of how thick the lithosphere is. They have helped us to narrow down the reasonable choices for interior structure, but interpretations vary depending on the models of interior composition and temperature. Many people are beginning to think that there is not a unique answer to the question of whether Venus has a thick or thin lithosphere. It may depend, to a larger extent than on Earth, on where and when you look.

If it seems that we are grasping at straws here, well, that may be true. The science of the Venusian surface does seem to be stuck in some sort of Newtonian quagmire where every idea is accompanied by an equal and opposite idea. At this point I don't know which is more confused—the crumpled, rifted, flooded, and folded Venusian crust, or the scientists who are trying to fit it all into a coherent global framework.

But we do have some good excuses. Here are three: First, it is no small task to figure out a world. *Magellan* has suddenly given us an almost overwhelming amount of new data. Venus is a large and complex planet. To borrow a phrase from Einstein and twist it: the most comprehensible thing about Venus is that it is so incomprehensible!*

---

*Einstein's famous remark is "The most incomprehensible thing about the universe is that it is so comprehensible."

Second, funding for planetary research continues to decline. In 1992 NASA started a three-year program to support post-*Magellan* studies of Venus. A lot of good research began under this Venus Data Analysis Program, or VDAP.* But VDAP was scaled back after one year and canceled after two. Some talented researchers have left planetary science for greener pastures. Third, although we finally have good surface and gravity maps, we still cannot study Venus with many of the tools we use to study Earth. Much hinges on the thickness of the lithosphere and on details of interior structure and convection. For Earth, we have a much better idea what lies inside. Here, we have a major advantage: if we want to know what's happening underground, we can just put an ear to the ground and listen.

INTRA-VENUS: WHAT LIES INSIDE?

You can look at the whole effort to understand the global tectonics of Venus as an attempt to connect the surface patterns observed by *Magellan* with structures and motions that lie hidden underground. But how can we hope to really know anything of what lies inside this other world, a world where even touching ground briefly and mapping the surface have proved so difficult? We have had to use a lot of educated guesswork, arguing largely by analogy with the Earth.

We are not completely clueless, however. We start with the assumption that Venus and Earth are basically made of the same stuff. Our current ideas about planet formation argue for this homogeneity. We know that the overall density of the planet is quite similar to Earth's, supporting the assumption of similar raw materials. We know more about the kinds of rocks at a few sites where Soviet landers did some experiments. We have hints about interior motions from the *Venera, Pioneer,* and *Magellan* maps. So we do know a thing or two. But if all we had for Earth was a similar collection of clues, we would be much more confused about the nature of its depths. Our most important source of knowledge about our own planet's insides comes simply from listening to the Earth.

---

*This included some important interdisciplinary projects that can fall through the cracks of the traditional funding structure, which divides planetary researchers up into Planetary Atmospheres and Planetary Geology.

How do you figure out what's inside anything when you can see only the outside wrapping? A package arrives in the mail and we shake it, listening and feeling for the response. When the Tin Man wanted to prove to Dorothy that he had no heart, he asked her to bang on his chest, and the hollow thud convinced her. When you knock methodically along a wall of your house and listen for the studs, you perform a similar exercise. In all these cases, sound reveals the density of the insides we can't see. We use this same type of exercise to listen for the interior density structure of our planet. Our ears for this purpose are seismic stations. Seismic waves are low-frequency sound waves that travel through the body of Earth, at speeds that depend on the density of the materials they pass through. By listening carefully, analyzing the pitch, timbre, intensity and rhythms of these planet waves, we can decipher much about Earth's interior structure.

We can directly observe only a very thin outer skin of the planet. Our deepest drill holes are less than ten miles deep, but the planet's radius is 3,955 miles. To sense the rest, we wait for earthquakes or set off explosions and listen to Earth's response. Using global networks of seismic stations, we have learned the basic spherical structure of our Earth—a solid nickel-iron core, a molten outer core, a heavy rock mantle, and a light rock crust. On a local scale, seismology reveals hidden faults, layers of sediments, and variations in the thickness of the crust and lithosphere. We do not have to debate the thickness of our lithosphere, because we can sense it with seismic data.* Correlating this underground probing with studies of surface structures is a very valuable tool for understanding Earth's geology.†

More recently, seismologists have been able to make three-dimensional images of some deep nonspherical structures in Earth. This "seismic tomography" is the geophysical equivalent of using ultrasound to image a fetus in the womb. Continent-sized bumps have been found on the core-mantle boundary, and the mantle is revealed as a lava lamp of rising hot plumes and sinking cold sheets of fluid rock. A particularly large hot blob is now rising up under Africa, and another

---

*We can also directly measure the heat flow at various locations of Earth's surface. Similar measurements for Venus would not be easy, but they would be a big help.

†I suppose this compensates somewhat for the fact that, compared with Venus, so much of Earth's raw bedrock lies hidden beneath sediments.

under the South Pacific. Global seismology is finally allowing us to see, or at least hear, the deep inner motions of our world.

We have never operated a seismic station on Venus. To do so under the spacecraft-melting and electronics-frying surface conditions would be no easy feat. Yet, ironically, one of our main tools for understanding the interior of Venus *is* seismology. Our models for the interior of Venus are all modified Earth models based on terrestrial seismic soundings. These analogies to Earth structure really save our hides. Without them we would be completely lost.

Perhaps this throws new light on our confusion about Venusian global tectonics. So much of what we know about Earth's interior and the way it relates to surface structures, we have learned from seismology. Yet on Venus we try to answer the same questions with *no* seismic data. Is this an exercise in futility? Only if we insist on finding the "right" answer, often a misguided goal in science, especially in a raw, cutting-edge field like planetary evolution. The confusion will gradually clear, and we will find closer approximations to the truth, which is all that scientists can really hope for.

THE WEATHER UNDERGROUND

Seismic observations not only reveal structures deep within Earth but also give us a snapshot of large-scale movements in the mantle. You can think of it as big, slow weather underground. These motions in Earth's mantle really are much like the weather we experience out here on the surface: the atmosphere is a heat engine in which sunlight drives the winds and convective motions of air redistribute the Sun's heat. These motions cause, and respond to, phase changes (between liquid, solid, and gas) that give us rain, snow, clouds, and evaporation.

Inside Earth is another engine, with convective motions of slowly flowing rock that redistribute heat from deep in the interior. The weather underground also has its phase changes, in this case changes in the crystal structure of rocks that occur at discrete levels of pressure in the interior, forming the rocky equivalent of cloud layers. One of these, known as the basalt-eclogite transition, occurs at a depth of forty miles below the ground. At this depth a descending slab of subducting basaltic crust is crushed into a more compact crystal structure, forming a rock known as eclogite. The heavier eclogite then helps to pull down the rest of the slab, furthering subduction. The presence of water in the rocks may be crucial for these phase transitions—yet another subtle way that water participates in plate tectonics. Sometimes the slow rock winds of Earth's interior weather send up a rising current of hot rock, inducing melt that rains upward to flood the surface with volcanic lava.

Living here on the surface of the Earth, we are perched between these two great convective systems that move on very different timescales, with human time somewhere in between. The weather above seems fast and chaotic and the weather below slow and predictable. But that's just a matter of perception. What we regard as predictable or chaotic depends on the timescale of our own experience.

Every convecting, chaotic system has a characteristic timescale of overturning motions, determined by the material properties of the convecting substance (air flows much faster than rock) and the spatial scale of the system (the mantle is thousands of times thicker than the atmosphere). The details of movement in Earth's atmosphere are somewhat predictable on the scale of a few days, but hopelessly random on the scale of a few weeks. This "chaos" explains why we'll never be able to predict the weather more than a few days in advance. The weather underground is also chaotic, but its motions are so much slower that it seems more predictable to us. I can write that a hot blob rising under Africa is causing a rift through Ethiopia, and it will still be true when this book is published and for millions of years to come. But by the time you learn that there is a low-pressure storm system off the coast of California, it may already be gone.

Remember the old *Star Trek* episode in which the *Enterprise* encounters a planet where time is sped up dramatically? The people there live out many hours for each second of "normal" time, and they are detectable only as passing insectlike buzzes in the air. If we could similarly speed up our conscious experience so that all of a human lifetime passed in a few seconds, the currents of the daily weather would seem as constant and dependable as the underground currents of the Earth seem in our normal time frame. We could study the structure of each weather front, publishing papers and holding meetings about each gust of wind, and they wouldn't go anywhere in the meantime. Conversely, if we could slow down our experience of time, so that tens of millions of years passed like moments, the internal convective breezes of Earth might seem as chaotic and ephemeral as the weather here on the surface appears to our normal time sense.

Venus also has weather above and weather below, heat-driven turbulent motions in the two great fluid mediums of gaseous atmosphere
and rock interior. In each of these realms, the style and character of the
motions differs from Earth's, in ways that are partly understood above
and at least suspected below. On Venus, with its thick atmosphere and
slow rotation, the motions of the lower atmosphere are well described as
giant convective Hadley cells (see Chapter 4). By contrast, Earth's rapid
rotation spins the atmosphere into cyclonic motions superimposed on a
background of Hadley circulation. Our ideas about the internal convection on Venus are more half-baked, since our data are so limited, but as
we have discussed, the convective motions there may be more tumultuous and organized on a smaller scale than Earth's.

Why should these planets have different interiors? A convecting
planetary interior is a nonlinear system, in which feedback allows new
qualities to emerge suddenly from a slight change in conditions. Models show that a small change in heat flow, perhaps from gradual overall
planetary cooling, can cause a sudden transition between the more
organized pattern of Earth, to the apparently more chaotic convective
behavior of Venus. Such an internal change in the weather 600 million
years ago might be responsible for the great transition seen on the surface of Venus. Could a similar rapid transition cause a global catastrophe in Earth's future?

LONG-LOST SISTER: EARTH'S TWIN RECONSIDERED

Much has been made of the "failure" of Venus to exhibit obvious Earth-
style plate tectonics. The *Magellan* findings have repeatedly been presented as revealing that Venus is a much more alien place than we had
expected, with no sign of the hoped-for Earthlike behavior on its surface. Almost every magazine article about the *Magellan* results pro-
motes or reflects this theme. I don't see it this way.

Venus clearly has her own style, but then so does Earth. Each is
unique in the solar system and surely, in some ways, in the universe.
Venus is not our identical twin. When they made Earth, they broke the
mold. There is no other. But Venus may be a fraternal twin. Born
together, under similar conditions, and growing up in slightly different locations, we two have gone our own ways to some extent. Upon
our first glimpse of her long hidden face, we are struck by superficial

differences. We think "She doesn't look a thing like me!" Later, we begin to notice the many traits that reveal a deep kinship, a close family resemblance that sets these two worlds apart from all others.

When *Magellan* was on its way to Venus, we were very much concerned with the question of whether Venus does or does not have plate tectonics. Perhaps now we are learning that a simple comparison using Earth as the standard is not the best way to think about Venus, and a simple yes-no question (does she or doesn't she?) is not the most meaningful one to ask about plate tectonics on another Earthlike planet.

Should we really expect to ever find another planet with plate tectonics exactly like Earth's? Or are there many possible styles of plate tectonics, one of which is found on Earth? The Venusian version may have a larger number of thinner plates and wider zones of deformation, or a more stop-and-go action than Earth's steady, well-lubricated drift. Perhaps we need a more pan-global conceptual framework for describing the styles in which planets rid themselves of internal heat and remake their surfaces. Someday we may be able to examine hundreds of terrestrial planets around our galaxy and derive a meaningful classification system of global tectonic styles. There may be hundreds of variations on the theme of plate tectonics. In this light, our expectation of possibly finding Earth-style plate tectonics on Venus may seem a bit like expecting to find human beings living elsewhere. Like a biosphere, a planetary system (with a convecting interior and a solid, mobile surface) is a complex, evolving entity with many possible outcomes. Historically, our discovery of Earth's plate tectonics is a very recent development, which has had a huge impact on our thinking about how Earth works. Given this recent revelation, we might naturally think of it as something a planet either has or doesn't have. But maybe you can have a little bit of it, or a version of it. Here on Earth, plate tectonics is *the answer*. Does that necessarily mean that it is the right question, when looking elsewhere?

When we look beyond the "disappointment" of learning that Venus doesn't have our homegrown version of plate tectonics, we can see many important similarities between the two planets. Although the details are different, if you squint your eyes a bit, Venus shares many of the attributes that otherwise would leave Earth an odd ball in the known universe. For one thing, Venus, like Earth, is a geologically active world. Very few places in the solar system have been able to make this claim for a long, long time. Discounting a couple of exotic moons orbiting the Jovian planets, all other planets have been pretty much dead, geologically speaking, for over a billion years. In the pre-Magellan era, Planetary geology, was largely a process of trying to reconstruct the distant past. We looked back to times when the surfaces of other planets had volcanic and tectonic activity which can be compared to current processes acting on Earth. Now we have found that there is another survivor nearby.

Almost all planets still bear the scars of the solar system's violent birth. The ancient terrains of Mars, the Moon, and Mercury are saturated with craters from the intense bombardment that marked planet formation's final phase (see Figure 5.19). As *Magellan* parted the clouds, we learned that Venus is the only other terrestrial planet where the activity of the surface has been intense and recent enough to hide all signs of this phase. Long after the other planets cooled and died, the restless, insomniac, tossing-and-turning interiors of Venus and Earth went on churning with their heat of formation and radioactive decay,

5.19 **Lunar Highlands:** *This heavily cratered area of the lunar highlands is typical of the oldest surfaces in the solar system, dating back to the massive bombardment that followed planet formation. Only Venus and Earth (in the inner solar system) have erased all signs of this phase.*

remaking their surfaces time and again, leaving no trace of the ancient pockmarked patches found on the skin of their planetary brethren. As on Earth, much has happened on Venus. It is obvious that our long-lost sister has been through a lot, as have we, and it will take a while for the whole story to come out.

FRESH AIR

It's not just the surface of Venus that is young and active like Earth's. The atmosphere may be quite youthful and actively evolving. We are still trying to figure out the exact composition of the lowest part of the atmosphere. And we have a lot to learn about the various sources and sinks that add and remove gases, and how fast they operate. But our work so far in this area suggests that the atmosphere of Venus is a dynamic place, with many vigorous cycles of activity that are constantly recycling and refreshing various gases.

Earth is an enormously complex system, with cycles and feedbacks operating on many different timescales. Through these cycles Earth itself breathes, cyclically exchanging matter and energy with the atmosphere. Much of the progress of "Earth system science" in the last two decades has been in working out the details of this web. Earth may be unique in our solar system in possessing these complex, self-regulating, chemical feedback cycles.

Or we may have company. In Chapter 3 I discussed the complex sulfur cycle of Venus, which has some intriguing similarities to Earth's carbon cycle. Venus probably has an active carbon cycle as well. Volcanoes supply CO_2, and carbon in the atmosphere takes many forms until minerals in the ground remove it from circulation. These rocks are eventually buried and remelted, giving their carbon back to the air. Unlike the carbon cycle on Earth, where most carbon is locked up in rocks and sediments, on Venus most carbon is likely in the air.

One reason we care so much about the carbon cycle on Earth, other than the fact that our own bodies are part of it, is that it dominates the climate evolution of our planet. Long before internal combustion engines, or clear-cutting of rain forests (or even rain forests), our planet's climate was evolving because of natural exchanges between CO_2 in the air and carbon reservoirs in sedimentary rocks and Earth's interior. Chapter 4 described some of the feedbacks that have contributed to this long-term evolution.

We are just starting to think about how similar feedbacks between the surface and atmosphere may be affecting the climate on Venus. Although the *Magellan* mission was mostly about the surface, it has caused us to rethink much of what we thought we knew about the atmosphere.

Before *Magellan*, we had no idea how volcanically active the surface was. Now we know that Venus has a young, active surface with fresh volcanic flows. Inspired by this, and recent refinements of atmospheric knowledge from the near-infrared windows (Chapter 3), we have begun constructing a new generation of atmospheric models that take into account the effects of ongoing volcanism. The preliminary results suggest that climate on Venus may result from a delicate balance created by the coevolution of the atmosphere, surface, and clouds.

The key players in the mighty Venusian greenhouse are CO_2, SO_2, H_2O, and the clouds themselves, all (including, ultimately, the clouds) probably being vented from volcanoes. These gases also probably participate in chemical reactions with surface minerals. We already discussed how the great heat probably makes chemical weathering an important phenomenon on Venus. This same effect probably means that surface minerals are involved in the evolution of the atmosphere, and the climate, of Venus.

Now, this is where things could get strange and complex, because not only the rates but the specific outcomes of these "surface-atmosphere reactions" are highly temperature dependent. If you have, say, SO_2 in contact with sulfur-containing minerals, and you change the temperature a bit, the chemical reaction changes as well, and the rocks will either suck in or pump out SO_2. The temperature at the surface *determines the amounts of these gases in the atmosphere*. Because of the greenhouse effect, however, the temperature at the surface *is determined by* the amounts of these gases. This creates the potential for some bizarre feedbacks. The surface of Venus may be actively participating in the evolving climate of the planet.

IS SISTER UNSTABLE?

Have you ever played, or at least heard of, the computer game "Sim-Earth"? This clever game lets you "play God" by altering various environmental factors and watching the effects on the evolving Earth system.

Like any computer model, the game is only as good as the equations that go into it. It is far too simple to be an accurate simulation of Earth's evolution, but it does give a feel for the way various components of our environment interact.

We have started to construct evolutionary climate models of Venus that include simulations of the surface-atmosphere reactions and the greenhouse effect. In a sense, we are trying to create a "Sim-Venus" that really works. Our early results seem to confirm that these complex feedbacks are important in determining the Venusian climate, but something about the results is troubling.

We keep finding that Venus is unstable.* This means that it could run away to much hotter or colder conditions with the smallest provocation. If anything happened to make the surface just a little hotter, the effect on the surface-atmosphere chemical reactions would be to pump more greenhouse gases into the atmosphere, and the planet would become still hotter. If something made the surface a little colder, greenhouse gases would be removed, furthering the cooling effect. In other words, surface-atmosphere reactions on Venus seem to create powerful positive feedbacks that quickly magnify any initial temperature change, causing either a runaway greenhouse (to conditions much hotter even than those found today) or a runaway cooling.

Why do I say that these results are troubling? Well, I don't believe that Venus really works this way. Natural systems generally don't hang around at unstable equilibrium points. Finding Venus poised at such an unstable state, just when we happen to come along and look, is like walking into a room and finding a pencil balanced on its tip. If you see this, something hidden is probably holding up the pencil. Most likely, something else is stabilizing the climate of Venus. There must be more going on, other *negative feedbacks* that lend stability to the Venusian climate system.

I think these early results tell us that more is happening on Venus than we have included in our models. Perhaps, as on Earth, the clouds have a stabilizing effect on the Venusian climate. After all, the clouds determine how much solar radiation reaches the surface, and how much infrared radiation makes it back out into space. The next thing

---

*Rather, I should say that our models are unstable. A planet will always be much more complex than any model.

we need to do with the evolutionary climate models is try to include the effects of evolving clouds. Another possible stabilizing influence is ongoing volcanoes. Maybe volcanic emissions of greenhouse gases are occurring at just the right rate to prevent a cooling episode.

So far, we are having trouble just understanding the stability of the current climate on Venus. During catastrophic resurfacing, when the rate of volcanism was much higher, the gases accompanying the huge volcanic lava flows would have made Venus much hotter. The problem is that we don't see how it would have returned to its current (comparatively) cool state.

Venus may have gone through some rather intense "global change" in the past. The average surface temperature could have been hotter, or cooler, by hundreds of degrees. This raises other interesting possibilities. Since temperature affects the viscosity of rocks, such intense climate changes could alter the overall behavior of the surface, and possibly even change the operation of Venusian "plate tectonics," (or whatever you want to call it). But the global tectonic system ultimately controls the rate of volcanism, and volcanic gases may be controlling the climate. You can see what I'm getting at. This is very speculative, but there might be feedbacks that, over long timescales, involve the whole planet, including clouds, atmosphere, surface, lithosphere, and even mantle and core.

We are finding out that Venus is a complex place that is difficult to simulate with simple models. This makes our task challenging, but it also makes the results potentially very rewarding, because, although Venus is clearly very different from Earth, it may rival Earth in complexity. Venus, like Earth, and like your body, may possess a complex interweaving of positive and negative feedbacks. The "Venus System" is certainly the closest thing around to the system of chemical cycles and evolving climate that defines the character of Earth. In the same way that the Venusian example may help us to generalize our ideas of planetary tectonics, Venus may be very valuable in helping us develop a general "planetary system science" of which "Earth system science" is a particular branch.

The importance of exploring Venus and understanding its climate is often presented as a cautionary parable of a world gone wrong, of what could happen here if we don't mend our foolish ways and stop this global warming. Like those antismoking advertisements that show us a heavy smoker's disgusting, blackened lungs and say "Don't let this happen to you," Venus is held up as an example of what we could do to our climate if we don't curb our addiction to fossil fuels. But this scare tactic does not really present the best rationale for studying Venus. We are not really in danger of triggering a runaway greenhouse that will boil our oceans and turn Earth into another Venus. We don't need to do anything nearly this extreme to make our planet very uncomfortable for us.

We are, however, dangerously ignorant about how planetary climate systems work. Studying other planets, and Venus in particular, is essential to remedy that ignorance. In reality, we should be more enticed than frightened by the Venusian example. Venus presents us with a wonderful opportunity. The lesson we can learn from the Venusian climate is not "Don't let this happen to your planet," but rather "Here is another nearby planet with a complex, evolving climate system; study it and you will achieve a more mature, less provincial, understanding of planetary climate." And that *could* save our hides.

BUT WHAT ABOUT MARS?

You may have read elsewhere that Mars is the most Earthlike of other worlds. At least two recent books about the solar system have chapter titles that read something like "Mars: The Planet Most Like Earth." Here I have been detailing all the remarkable ways in which Venus is uniquely like Earth. What gives? Obviously, somebody must be wrong.

Actually, both points of view are correct. Both Mars and Venus are most like Earth in some ways. It depends on what you're looking for. On Mars, some of the *surface conditions* are a lot closer to conditions on Earth. In particular, its range of temperatures is a lot closer to, and even overlaps, Earth's. On the hottest days of equatorial Martian summer, it sometimes reaches 32 degrees Fahrenheit. The difference in surface pressure is also much less. In these ways, Mars would be a less hostile environment for us to explore in person. A human Mars suit is easier to design than a Venus suit, and it is less of a challenge to build machines that can work on the Martian surface. Mars's twenty-four-hour day, and the sight of the Sun and stars in the sky (which you could never see from the surface of Venus), would also be comforting. So it is easier to imagine ourselves going there to explore and live.

Mars also shows tantalizing signs of water running on its surface in the distant past. This has fueled much speculation about a possibly more Earthlike past on the red planet, with a warmer and wetter climate and (just possibly) life. The tantalizing recent discovery of possible signs of ancient life in a meteorite from Mars certainly encourages such speculation. Venus, too, most likely had liquid water long ago, but she does not wear her ancient history on the surface the way Mars does.

The restless interior and surface activity of Venus has wiped out traces of all but the most recent history. It is in just this important aspect that Venus resembles Earth more closely. Geologically speaking, Mars is dead, while Venus and Earth live on. As befits a planet of its smaller size, the interior of Mars has cooled to the point where it no longer has volcanic or tectonic activity on the surface. By contrast, Venus and Earth are warm to the touch, alive and kicking. The vigorous, dynamic surfaces of Venus and Earth are reflected in the complex chemical cycles of their atmospheres. Thus Venus system science may prove to be a more interesting analog to Earth system science than a similar study of Mars.

I don't mean to imply that Mars is not an incredibly interesting place worthy of our exploration and perhaps even eventual habitation.* Mars and Venus are both tremendously valuable places for us to study, and they can be useful to us in different ways. Venus may prove most valuable for learning to think about the Earth, and Mars for thinking about our possible future and destiny among the stars. Mars may help us find clues to the origin of organic life. For the purposes of taking the next human steps into the rest of the big, wide universe, the place to look is Mars. But we will learn most about planetary formation and evolution by studying and comparing all three of these worlds. So, I am not at all advocating that we ignore Mars, only that we pay attention to Venus, and the stories she can tell us about our world, its lost past and its possible futures. After all Venus and Mars together are our closest family, our next of kin, our siblings. There is no need for rivalry: Let's get to know them both.

---

*I know I'd better watch what I say here. There are agents of the *Mars Underground* lurking everywhere.

EARTH REVEALED

In Werner Herzog's movie *The Mystery of Kaspar Hauser*, the title character spends his childhood and adolescence locked in a basement, in complete isolation from human society. When freed from this solitary confinement, he has physically grown to adulthood but must learn from scratch all about the external world and learn to see himself in a societal context. Through his eyes we see human interactions from a fresh and often surprising new perspective. As a species struggling to find our place on our planet, we have spent our childhood and adolescence in similar seclusion. For our first 4.5 billion years as a planet, 4 billion years as a biosphere, 3 million years as a species, and few thousand years as a "civilization," we have had only hints that similar worlds might exist. But self-knowledge can proceed only so far in isolation. Now, with the parting of the clouds of Venus, we have been let out of the basement.* We can examine a true peer and see our planet in the light of knowledge of another.

By sending us back to the drawing board to reconsider basic questions of geology, post-*Magellan* Venus studies will have a healthy effect on the earth sciences. I put this in the future tense because despite some initial tremors, I think most of the ripple effect has not happened yet. The new questions about geophysics and plate tectonics raised by *Magellan* have not fully percolated out into the community of terrestrial geologists. But they will, because thinking about Venus makes us keenly aware of weaknesses in our understanding of Earth. For example, Venus can help us think about long-term evolution, by providing glimpses of possible once and future Earths. Our attempts to determine whether Venus has plate tectonics makes us question how (and when) plate tectonics got started here. Subduction on Earth may be self-propagating once it gets going, as descending slabs pull the plates along. But how do you get slabs down to that depth in the first place? What primed this pump? Some scientists are trying to model the initiation of subduction on Venus as they attempt to understand possible episodes of catastrophic resurfacing. These same geophysicists are now training their models on the early Earth, seeking the origins of plate tectonics.

---

*The ultimate analogy, of course, would be finding another "intelligent" species we could talk to. In the absence of this, having another similar planet to study can greatly enhance our perspective on the home world.

Some researchers have suggested that Venus resembles Earth before the time of plate tectonics. At some point in the past, Earth must have had a higher heat flow than it does at present. This would mean a thinner lithosphere and less rigid plates. Perhaps Earth's first global tectonic system was characterized by crumply plates, coronae, and an absence of subduction, similar to Venus.

As with many ideas in this frontier field, an opposite idea also holds sway. Venus has been relatively quiescent for the last 600 million years, with most of the surface activity concentrated at a relatively small number of hot spots over mantle plumes. One idea for the future evolution of Earth is that eventually, as the interior cools, mantle convection will become less vigorous and plate motion will cease. The remaining heat flow and surface deformation will be dominated by plumes. So we don't know if Venus more resembles our future or our past, or if neither one is a good analogy. But these comparisons are fueling new speculation about Earth's very long term evolution.

One persistent idea about Venusian tectonics is that surface recycling on Venus may proceed in fits and starts, compared with Earth's relatively steady plate motion. Yet, recently we have learned that plate tectonics, the "uniformitarian" mechanism by which Earth sheds its excess heat and remakes its surface, actually proceeds at least in part through a series of sudden events. The rates of plate motion and production of new crust on Earth have gone through some episodic changes. It seems that every few hundred million years a tremendous blob of material, like an immense hot-air balloon of superheated rock, rises from Earth's core-mantle boundary. These "superplumes" ascend two thousand miles through the relatively fluid mantle and hit the base of Earth's rigid lithosphere, melting it on a gigantic scale and flooding vast areas of the seafloor (or occasionally a continent) with torrents of fresh lava for tens of millions of years. These major hiccups in Earth's internal workings have all kinds of other global effects, including changes in sea level and increased CO_2 in the atmosphere from the outpouring of volcanic gas. Superplumes cause rapid subduction, as oceanic crust plunges back into the mantle to make way for the pulse of newly created crust.

One hundred twenty million years ago, during the most recent of these events, the "Mid-Cretaceous superplume episode," Earth's crust production suddenly doubled and then slowly declined over the next

70 million years. The huge complex of volcanic landforms formed during this event still dominates the geography of the western Pacific sea floor. The CO_2 released warmed Earth by about 20 degrees Fahrenheit. Oceanic plankton growing in these warm waters ultimately produced about half of our present reserves of oil, fueling our current orgy of internal combustion that threatens yet another global warming event.* The expansion of the crust from this outburst caused rapid subduction along the eastern margin of the Pacific, and the subducted crust quickly melted, erupting into new volcanic mountain ranges. Thus, California's Sierra Nevadas and the Andes in South America also owe their origin to this most recent superplume event.

We have a great deal to learn about why Earth goes through such changes in activity level. Maybe the apparent global oscillations on Venus represent an extreme of the same type of behavior. Perhaps a comparison of the two will allow us to better understand why terrestrial planets experience such oscillations.

Ultimately, both Earth and Venus must be in decline. All planets are cooling off and becoming less active: look what happened to Mars. It's only a matter of time. The overall rate of heat flow, vigor of mantle convection, and amount of surface activity will decrease as the lithosphere thickens and the interior cools. Both planets experience oscillations of increasing and decreasing activity superimposed on this overall decline. Certainly, over the last 600 million years, Venus has had a less active surface than Earth. During this same period the overall rate of plate motion on Earth has gone up and down by about a factor of two. Venus may experience more intense episodes of both increased and decreased activity. Sister Venus seems to have a bipolar personality, as compared with our more steady demeanor. But both are aging, and it is not clear which one is more well preserved in the sense of retaining more interior heat, which for planets is the elixir of youth.

One weakness of current plate tectonic theory is that tectonic structures *within continents* are not well understood. The theory does a good job of explaining the geological activity along the boundaries between oceanic plates, but it doesn't explain the large, widely distrib-

---

*The carbon atoms that warmed Earth as CO_2 100 million years ago were incorporated into the bodies of plankton using solar energy and buried in organic sediments. The very same atoms are now being reunited with oxygen in our cars, factories, and forest fires, to warm Earth yet again.

uted areas of deformation (mountain building and faulting) within the continents. The basin and range area of the western United States is one of these zones. It has been stretched out over hundreds of miles into a repeating pattern of great basins and mountain ranges, with faults in between. Unlike the narrow zones of activity at plate boundaries, which are typically a few tens of miles wide, zones of deformation on the continents are often hundreds of miles wide. Geophysicists do not agree on how best to describe the deforming of the continents, but it is probably not due to the interaction of rigid plates. Some describe the continents as broken up into many "microplates" that slide against one another in these broad zones. Others see the continental crust as a more fluid medium that stretches and flows throughout these zones.

For this quest, Venus may be instructive. Indeed, it seems that the whole surface of Venus may behave more like Earth's continental crust than its oceanic crust: blocks of relatively stable material are separated by wide zones of active deformation. I would not be surprised if, over the long run, continued studies of Venusian tectonics helped us solve the riddles of Earth's continents.

These are some of the ways that studying Venus may help us deepen our understanding of global-scale tectonics on Earth. On a smaller scale, studying individual features in the "naked geology" environment of Venus, where fresh bedrock is not quickly covered by sediments, may give us new insights into the origin of some of Earth's geological forms.

Turning back toward Earth with fresh eyes, we begin looking for a deep and unified understanding of the many differences between these worlds. As we try to make sense of the global portrait of Venus's crumpled volcanic surface that *Magellan* has presented to us, and search for unifying explanations that will render the big picture explicable, we keep finding ourselves returning to water as the source of understanding.

THE IMPORTANCE OF BEING WET

Venus may have "rivers" and "snow," but she has no water, unless you count the shockingly small number of water molecules we detect in the atmosphere. It is somewhat ironic to find our twin planetary sister in the grip of such extreme drought, because Earth is a water planet. No,

Earth is *the* water planet. Water has gotten into Earth's pores and bones. It is not confined to the oceans, lakes, and rivers that form three-quarters of our planet's surface. Ninety-seven percent of Earth's fresh water is underground, seeping through rocks and soil and running slowly to the sea in hidden rivers. An unknown amount of water, possibly more than in the oceans, hides in the mantle. Earth is a soggy planet, saturated through and through. Water flows ceaselessly: evaporating from ocean to cloud; raining onto the ground; infiltrating, exposing, dissolving, and corroding rocks; and then running in creeks, streams, and rivers back to the ocean. The presence of water shapes every aspect of Earth's being: her unique atmosphere, her ubiquitous biosphere, her rapid erosion, the mechanics of her global plate-tectonics system, and even the properties of her mantle and deep interior—all are shaped by water.

Water is also perhaps the key ingredient in the *aesthetics* of Earth. Watery places, and places where water has been and left its signature, form some of Earth's best and most beautiful sights, her Scenic Overlooks and Points of Geologic Interest.

We never fully appreciated just how pervasive the role of water on Earth is until we had our recently clarified views of the atmosphere and surface of Venus. The overall character of the Venusian surface at every scale is defined by familiar rock-forming processes, unhassled by the familiar rock-destroying tactics of water.

On Earth, water facilitates the weathering of rocks, ceaselessly grinding down any mountain or rock bold enough to assert itself skyward. Weathering of rocks also affects our atmosphere and climate, constantly removing CO_2. If the young Venus, just a bit more influenced by the Sun than Earth was, lost her water early in a "runaway greenhouse," then she may also have lost her ability to easily cool herself by removing CO_2 from the atmosphere.

If Venus is so nearby, so similar in size, and made out of the same stuff with the same initial heating, why doesn't she exhibit the obvious signs of plate tectonics, at least as we know them? Why is her surface not divided neatly into basaltic ocean crust and granitic continental crust, like ours? And why is her climate so "extreme" and "inhospitable"?* It may all come down to water.

*Any Venusians could easily ask the same questions of Earth.

Earth's unique hydrosphere may contribute to our unique style of plate tectonics in several ways. We've known for a long time that water is *involved* in plate tectonics. But when we look at Venus, then back at Earth, we ask, is it *essential*? Earth's plate-tectonics engine may be lubricated with water. Water-driven phase transitions in our mantle may be essential for subduction, and this may be what keeps Earth's tectonic conveyer belt running. Water may be necessary for the chemical reactions that make the light rocks, like granite, which form Earth's continental crust. The upper crust of Venus may be greatly stiffened by its dryness, and this may help to hold up her highest mountains. This could also prevent the crust from participating in our kind of plate tectonics.

It may be that the existence of Earth's unusual plate tectonics, rather than the possible lack of it on Venus, is the real mystery. How is plate tectonics so well coordinated that it maintains just the right flow of heat to avoid an episodic evolution? This is not just an arcane question of geophysics. So many of Earth's other qualities—its atmosphere, climate, hydrosphere, interior, and so forth—are intimately tied to plate tectonics that plate tectonics may be directly tied to Earth's habitability for our kind of life. If this overriding aspect of Earth's evolution is very unusual, very rare in the universe, then this may bear on the likelihood of finding other planets with Earth-style life.

Our earliest biological roots are to be found in the waters of the young Earth. According to our current ideas about atmospheric evolution, it seems that Earth's early atmosphere may have been a thick blanket dominated by carbon dioxide, much like the modern-day atmosphere of Venus. Venus may be the place in our solar system that most resembles that early Earth when life was gaining its first toehold, except for one crucial fact: early Earth was wet, and Venus is dry. This is no minor distinction. On Earth water and life are inextricably linked. Like Christopher Robin and Pooh, wherever one goes, the other goes too. In a sense, life is a specialized organ of Earth's ocean, an enterprising bit of it that has organized and figured out a way to explore the outer space of land and air. We still carry the ocean in each of our cells. We are walking sacks of salty seawater with some dissolved organic compounds.

Life has been around for most of the age of Earth. Earth and life have grown up together as blood brothers, with water the blood that binds them. Has life contributed in a dominant way to Earth's unique environment, or has life been a more passive partner, merely adapting to and benefiting from the changing world? In recent years the debate about how central a role life has played in Earth's evolution has heated up, largely due to the Gaia hypothesis. This chicken-and-egg question is a difficult one.

It is hard to imagine Earth without water and life. But perhaps Venus can help. At least in some ways, Venus can be seen as a "control" for the "effect of life on Earth" experiment. You take two similar planets, add bugs to one, wait 4 billion years, and watch what happens.

But wait: do we know for sure that Venus is, and has always been, lifeless? Thinking of the contrast between our living Earth and its apparently lifeless sister planet allows us the opportunity to ponder the essential qualities of, necessary conditions for, and universal features of life. Can we answer these questions using our present knowledge?

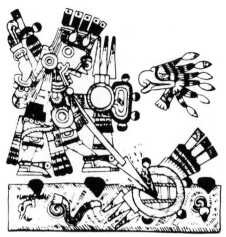

6

life on venus: a barren world?

The totality of life is merely a fancy kind of rust, afflicting the surfaces
of certain lukewarm, minor planets.
—H.J. MULLER, *Life* (1955)

NEW WAYS OF LOOKING AT LIFE

Is there life on Venus? Could there be life on Venus? The standard
answers are "No" and "NO!" Venus is usually dismissed in a paragraph
or two before an extensive discussion of the prospects for life on Mars,
the icy Moons Europa and Titan, and Earthlike planets elsewhere in the
universe. Where life is concerned, Venus is consistently voted "least likely
to succeed." In my opinion, this quick dismissal is not justified. It pre-
supposes knowledge of the universal nature of life and the general characteristics of inhabited planets—knowledge that we do not yet possess.

Life is usually assumed to require organic molecules* dissolved in
liquid water. Discussions of the habitability of other places in the universe have focused almost exclusively on the likelihood of finding planets with climate and atmospheric conditions that are "just right" for us.

*Complex molecules dominated by carbon.

293

By this standard model, Venus is obviously sterile because it is too hot and dry there. But I think it may be premature to declare Venus off-limits to life. I know that some may find this viewpoint ridiculous or even irresponsible; it could get me in hot water with some of my colleagues, but that's okay. Hot water, as a symbol for domains that are off-limits to life, is a very appropriate metaphor here.* In our present state of ignorance we should avoid all dogma in exobiology (the study of possible life beyond Earth).

Part of the problem is that life is difficult to define, so how do we go about looking for it on other planets? Life is like a standing wave in a swiftly flowing stream—a stable structure through which matter and energy flow. If we are going to look for it around the universe, however, we need a more specific definition. Usually we define it as something like "a self-propagating chemical system that adapts to its environment." Two recent scientific movements suggest refinements to the definition, putting some new spin on our ideas about habitable places elsewhere. These two new developments are Gaia and *complexity*.

I've already discussed Gaia in connection with life's possible role in Earth's evolution, but this perspective is also valuable for thinking about life elsewhere and how we might recognize it. Gaia has its roots in the ideas of a remarkable Russian geochemist, Vladimir Vernadsky, who described the "biosphere" as that part of Earth transformed by life. He defined life as a planetary property, as the way Earth's surface responds to sunlight, covering itself with blue-green algae, forests, prairies, and eventually parking lots and shopping malls.† "The Earth is literally covered with an uninterrupted film of living matter," Vernadsky wrote in 1926. He described how this zone of life had modified the planet, facilitating further development of life, and how life's participation in numerous geochemical cycles dominated the appearance, structure, and chemical state of Earth's surface and atmosphere.

More recently the Gaians have written of "geophysiology," describing mechanisms of self-regulation by which life may be actively controlling the climate and other conditions on Earth. Some of the details are still quite controversial. In particular, the extent to which life

---

*Or maybe it will just be given an icy reception. Equally appropriate.

†Fortunately for him, I don't think Vernadsky lived to see the advent of the modern mall. He died in 1945.

is really in control of Earth's overall environment is subject to continued debate. But the Gaia hypothesis has motivated a lot of good research into the intricacies of life's role on our planet.

It cannot be denied that life has fundamentally altered the appearance, atmosphere, climate, and chemical composition of the Earth. Our world is awash in life. Interacting communities of organisms create one another's environments, and this "worldwide web" shapes and is shaped by Earth. In a very real and tangible way, then, Earth—all of it—is part of this circulating, self-perpetuating, complex, evolving, breathing, growing, dying, talking, and dreaming thing called life.

The biosphere is a feature unique to our planet, so far as we know (which isn't very far at all). Life has a hold on Earth and may never give it up, until the Sun expands and consumes us 5 billion years from now. By then we may be long gone, watching the Sun's dying pyrotechnics from a safe distance.

Life leaves its traces even in the "nonliving" parts of Earth. Everywhere we find patches of chemical and physical complexity dropped by life in its haste. Life is so full of life that it occasionally spills it on the ground, leaving distinctive signatures of isotopes and chemicals, or structural patterns and rhythms, for us to find. Should we find similar signatures and patterns on other inhabited worlds?

According to the Gaian view, life is a process that, once started, becomes intimately involved in the later evolution of its home planet. Any inhabited planet should be brimming over with life, and life should have affected it physically in ways we might observe.

This overall concept has merit—but do we know enough to make specific predictions? The Gaians have made a nice first attempt to define the properties of inhabited planets. Their criteria, though, are of necessity heavily based on the example of Earth. They claim that we can tell whether a planet is inhabited from the composition of its atmosphere. Inhabited worlds have atmospheres that are out of equilibrium.* Equilibrium equals death. They note how Earth's lively, cycling atmosphere, far from equilibrium, is deeply related to the planet's capacity to support life and the effect that life has on the planet. This is contrasted with the "dead" worlds of Mars and Venus,

*See Chapter 4 for a discussion of the meaning of equilibrium.

which they portray as static, in equilibrium, with atmospheres that are "just sitting there." This may be true of Mars, but such a view of Venus is outdated. Like Earth, Venus seems to possess lively chemical flows and cycles that may actively maintain the atmosphere and clouds.

One problem with the "disequilibrium" criterion for inhabited worlds is that it is vague. The atmosphere of Venus is definitely not in equilibrium. It is closer to equilibrium than Earth's, but how far from equilibrium must an atmosphere be before it is considered alive?

Another problem is that there are many nonbiological sources of atmospheric disequilibrium, such as lightning and ultraviolet radiation, even on the few planets we have studied. Furthermore, not all life creates chemical disequilibrium. On Earth, plants do, but animals do the opposite: they eat disequilibrium and excrete equilibrium. With every breath you take you inhale oxygen and exhale carbon dioxide, bringing Earth's atmosphere a bit closer to an equilibrium state. Life giveth disequilibrium, and life takes disequilibrium away. So do we really know what to look for? Shouldn't an atmosphere *suspiciously close to equilibrium* be just as likely to signify life?

In fact, the atmospheres of Venus and Mars are both "suspiciously" close to equilibrium. They are dominated by CO_2, yet in their upper regions ultraviolet light naturally breaks up CO_2, creating CO and O. Given this, atmospheric chemists have had some difficulty figuring out why the CO_2 atmospheres of these planets are stable. We would expect both planets to have much more CO and O_2 and less CO_2. Is this a sign of life on these worlds? Are there creatures that breathe in CO and O_2 and make CO_2? Probably not. We think that we've found natural chemical pathways that reunite the broken CO_2 molecules.

This example illustrates why inferring the presence of life from "strange" atmospheres is difficult. We don't know enough about nonbiological or biological atmospheric processes to make sweeping statements. However, it's good that people are starting to think along the lines of a biosphere as something with planetary-scale properties. This helps us think about what properties inhabited planets may have and how they may differ from lifeless worlds.

COMPLEXITY: WHAT'S THE BIG IDEA?

Matter awoke and organized itself. The flame gave way to music.
—HUBERT REEVES, *Atoms of Silence*

You've heard of the second law of thermodynamics: Entropy always increases. Things fall apart. If you don't keep cleaning your house, it will become hopelessly messy. All the king's horses and men can't put Humpty-Dumpty together again.

But how did Humpty get put together in the first place, if the universe always tends toward disorder? There are local tendencies toward increasing complexity and order that do not violate the second law, because it applies only to closed systems. If entropy (disorder) is lowered here, the universe pays for it somewhere else.* In our open system, Earth, life feeds off of the Sun. It takes energy to make ordered structures like us.

Actually, it takes a *flow* of energy or matter to keep an engine, a life form, or a biosphere running. As we discussed in Chapter 3, flows always create differences, or *gradients*,† a simple form of order. An engine or a life form feeds off a gradient in heat or chemical energy and always acts to reduce that difference. Without a continual input of new energy from outside, life forms and machines will erase all gradients, run down, and die, producing a state of minimum order and maximum entropy where nothing changes. It can also be described as a state of complete equilibrium. Pretty boring, if you ask me.

The universe tends toward death and disorder. It's the law. But look out of any window or down at your toes. Our world is fantastically ordered and alive. Somehow, on the surface of Earth, the flow of solar energy is transformed into the fantastically complex ordered system of the biosphere. Why here?

In some circumstances, patterns seem to emerge spontaneously from the flow of energy. The effort to understand how order comes from chaos has led to the creation of a field known as nonequilibrium thermodynamics. It was pioneered several decades ago by Nobel laureate Ilya Prigogine, who coined the term *dissipative structures* to describe the forms that arise spontaneously in flows of matter and energy, dissipating energy and creating order along the way.

A simple dissipative structure, one we have seen many times in this book, is a convection cell, in which heat flowing through a system

---

*The total entropy of the solar system increases every day, mostly because the Sun increases its entropy as it radiates energy.

†Recall the example of a gradient of salinity across the zone where rivers of fresh water flow into, and mix with, the salty sea.

organizes itself into discrete patterns of upwelling and downwelling matter. This organization can be very simple or more complex. The weather patterns in planetary atmospheres are dissipative structures created by the flow of solar heat. The flow of heat out of planetary interiors also results in the spontaneous organization of convection cells. Whirlpools and tornadoes are dissipative structures, pockets of order created within energetic flows.

Nonequilibrium thermodynamics also helps explain how the fantastic complexity of life on Earth can have arisen in a universe that tends toward disorder. Convection cells are the simplest example of dissipative structures; organisms and biospheres, the most complex. This tendency of matter to use flows of energy to form complex structures, in apparent local defiance of the universal second law, has also been called *self-organization.*

Have you heard any of the ruckus surrounding the new scientific field of *complexity theory?* It is the mathematical study of self-organization, a direct intellectual descendant of nonequilibrium thermodynamics, that was born with the advent of computers big and fast enough to simulate some of the complex processes in the real world.

The complexity theorists use computer simulations to mimic the emergence of order in natural systems, trying to develop an understanding of the principles governing the "emergent properties" that arise from matter in certain circumstances. I say "trying" because no one has yet developed an overall theory. But they have demonstrated intriguing similarities in the properties of many different kinds of complex emergent systems, including life forms, ecosystems, and societies.

We've all observed emergent properties. Sometimes a new behavior arises from interactions among the elements of a system—a new quality that resides more in the patterns of mutual interaction between the components than it does in the components themselves. One example is a phase change. When ice freezes or melts, the individual water molecules remain the same but their interactions change. Emergent properties can arise in groups of human beings, too. When a basketball team suddenly gets "momentum," it is something that happens to the team collectively and can't quite be explained by, or reduced to, the actions of individual players. Suddenly their passes are connecting, they are hitting the basket. They are on fire, as a team.

The scientists involved in the complexity movement are attempting to arrive at a new way of describing the emergence of biological order. They see life as an emergent phenomenon that arose on Earth because of the right conditions, involving flows of matter and energy and the availability of chemical building blocks able to tap the potential for order. They may be laying the groundwork for a theory of life that does not depend on the details of conditions on Earth, a "physics and math of evolution and metabolism."

Complexity does not threaten to replace Darwinian natural selection or render it obsolete, but it may augment it in ways that could throw new light on the marvel we call evolution. Darwinians speak of the "blind watchmaker" of evolution, which selects from completely random variations over long periods of time. The "mistakes" in the genetic code that enhance survival naturally tend to propagate themselves, and thus life slowly evolves. But some complexity theorists are now questioning whether the watchmaker is truly blind. They suggest that since nature seems to spontaneously seek order in some situations, this emergent behavior might be behind a lot of biological innovation. In their view, natural selection chooses from among a menu of spontaneously ordered options. Maybe the watchmaker of evolution is not completely blind, or is at least guided by some pattern-forming habits built into this universe.

That may explain why so many biological shapes are *fractals*. Fractals are forms with self-similarity across scale, that is, shapes in which the smaller parts are the same shape as the whole. Some examples are branching trees, spiraling seashells, and your nervous and circulatory systems. The most famous fractal of all is the Mandelbrot set—you know, those baroque, Day-Glo,* Dr.-Seuss-on-mushrooms designs you've seen on T-shirts and in music videos. These infinitely intricate patterns are all created from a simple mathematical equation, repeated over and over again on a fast computer.

The Mandelbrot set and other computer generated fractals often seem remarkably familiar. You watch fractal videos and think "There's a sea horse" or "There's a line of marching tigers." Why do the shapes of

---

*There is nothing inherently colorful about these patterns. They are just collections of numbers being represented with an arbitrary color scheme.

living things on Earth seem to have so much in common with these mathematically generated forms?* Fractals are a kind of geometrical complexity. Complexity theorists have found that many emergent self-organizing systems create fractal shapes.

Fractals are made by simple equations in *iterative* computer programs: the product of one round of calculation is used as the input for the next step. This is the mathematical equivalent of feedback, of a nonlinear process. Probably nature likes to make fractals because many natural growth processes are iterative. Somewhere their "instructions" contain the command "Go back to the previous step and repeat until done." For whatever reason, nature seems to love to assemble itself into fractals.

When you start looking for fractals on Earth, you see them everywhere. You see common geometrical forms in life from the micro- to the macroscopic scales. The same shapes appear in life made from very different materials, on land and in the sea. Fractal forms seem to arise from basic physical constraints on growth processes, universal physical realities. Whatever generates fractals may be deeply embedded in the rules of life, quite possibly deeply enough to transcend worlds. When we do finally meet the aliens, maybe they won't look completely unfamiliar! Anyone who has spent time looking at the Mandelbrot set might respond to the first pictures of complex alien life by saying, "Oh, I've seen *that* one before!" (See Figure 6.1.)

One possible answer to "What is life?" is "I don't know, but we'll know it when we see it." We might look for life on other worlds by examining photographs for well-developed fractals. This search strategy assumes no particular biological mechanism, which is good given our ignorance and nature's inventiveness.

Of course, there are geological forms, like branching rivers, that look fractal but are not necessarily related to life. Complexity theory would need to advance to the point where we can reliably predict which kinds of fractals are biological before we could make use of this strategy. Here is a prediction: We will find fractal forms, and they will seem familiar to us, when we find complex life on other worlds.[†]

---

*Some nonliving natural processes make fractals, too, most notably erosive structures, like branching rivers. It is interesting that these forms usually involve some of the same conditions that make life arise, like flowing water. Is this a coincidence?

[†]I love this kind of prediction. I can be proved right *or* wrong only by a fantastic discovery.

6.1 Universal Forms?: *These three fractal shapes are (left) a photograph of plant life at the Denver Botanical Gardens (Tory Read), (center) a drawing of a sea-cucumber (Ernst Haeckel), and (right) a computer-generated fractal, a mathematical abstraction. Will life forms on other worlds exhibit similar, and thus recognizable, shapes?*

Critics attack complexity theorists because they don't (yet) have a unified theory encompassing all the bits and pieces of this new research direction. They do, however, have a lot of intriguing results that are worth getting excited over, even if no one knows quite where it is all leading. Complexity represents a new way of looking at nature that doesn't quite fit the rules of science as it has been done. For example, it is inherently nonreductionist, focusing instead on emergence. One objection is that complexity science is not predictive—at least not yet, in the way that science has required itself to be. Maybe in some sense it is not even science but natural philosophy. In any case, it is an important new way of looking at things that have traditionally been the domain of science. Complexity seems able to describe a surprisingly wide range of natural phenomena underneath one emerging branch of theory. We could miss out on something important if we dismiss it rhetorically because it provides no unified, predictive theory or because it does not fit the old rules.*

Is life implicit in the laws of physics?† Does the universe have an inherent capacity or a tendency to become alive, given the right conditions? Many people have said that it does, that life has been *in* the

---

*This dialog on the value of complexity is interesting to watch because it is a battle of cultures or even world-views *within* science. The debate is not just about whether certain theories are right, but about a way of looking at the world, and even the boundaries and definition of scientific thinking.

†Some have said that complexity implies that we are "expected." I like "implicit" better than "expected," which implies that someone has been expecting us.

universe from the word *bang*. We cannot say for sure, at least given what we know now. We can poke around the edges of this question with science, but it is also fair game for intuition and faith. I feel fairly certain that the universe wants to be alive, that the arrival of consciousness here, and probably elsewhere, is inevitable. We are the universe waking up. I do not have any strictly scientific arguments to support this, but everything I know and have seen of this universe, informed and enriched by all my scientific studies, as well as direct experience, contributes to this viewpoint. I prefer not to believe that we are the only conduits for consciousness in the entire universe, not because I have any evidence to the contrary but because the notion seems absurd to me. It's largely a matter of aesthetic preference. One of the exciting implications, for me, about complexity theory is that it may represent the beginnings of a quantitative theory that could address this question.

The current paradigm of exobiology has existed for about forty years or so: life is carbon based, made by chemical evolution of organics in water. Earthlike environments are the only places where life can really be expected to flourish. You have to wonder, though, about any theory that concludes that our kind of life is the only kind, and that planets just like ours are uniquely qualified to become alive. If complexity theorists can develop a general theory of living systems, maybe we will have criteria less dependent on our experience within the biology of only one world.

Perhaps a combination of ideas from Gaia theory and complexity theory can help us to think more generally about planets as environments for life. Gaia encourages us to think of a biosphere as a global property that certain planets might possess. Complexity suggests that life emerges spontaneously amid the right kinds of energy and matter flow, and where conditions are suitable for stable structures to form and last long enough to perpetuate themselves.

What would make one planet better than another as a place for a biosphere to grow? Before we plant a garden, we till the soil, turning it over to ensure that nutrients are brought up from below and air is mixed in from above. A planet like Earth tills its own soil, with its constantly cycling interior, surface, and atmosphere. This activity keeps nutrients and energy sources constantly available, making the surface of Earth a fertile place, a good niche for life. Planets like this may be likely to form biospheres as a natural consequence of their physical evolution. Perhaps an active, overturning surface, and an atmosphere with dynamic chemical cycles, will prove to be good qualities for living worlds.

Even if you removed life, Earth might still seem like a living system, with its complex interweaving feedback cycles of matter and energy and many chemical reservoirs interacting and regulating each other on different timescales. Living beings are constantly renewing themselves, and living planets may do the same. Our skin seems permanent, but no cell lasts more than two weeks. Our bones seem solid but are always being deposited and reabsorbed, like an active planetary surface. So, just maybe this is what's really required for life. Maybe a planet that is geologically "alive" is more likely to be biologically alive. Maybe a living planet will always "look alive," in the sense that Earth does, with a constantly self-renewing surface and atmosphere.

Venus has an active surface and interior and a lively atmosphere with complex chemical cycles that perpetuate gradients of matter and energy. In theory, this kind of disequilibrium environment could feed a steady supply of nutrients and energy to any creatures crafty enough to evolve in the Venusian environment.

Of course, there is a serious objection to this. Neither organic (carbon) compounds nor liquid water can exist at the temperature of the Venusian surface. But how sure can we be that this rules out life?

LOVE THAT DIRTY WATER

The universe is not only queerer than we imagine, but it is queerer than we can imagine.
—J. B. S. HALDANE

First let me say that I am a big fan of carbon-based life. Some of my best friends are carbon-based. Carbon and water are two substances that each have incredible properties on their own. In combination they do something completely different, something that neither could do on its own. Carbon, with some oxygen, nitrogen, and a few other elements mixed in, serves as the universal template, the flexible yet solid Lego blocks of life that can build up an endless variety of huge and complex molecules. Water is the universal solvent. Dissolved in water, carbon molecules are free to flop around, twist themselves into complex shapes, and interact with one another in the intricate dance that we call life. Something magical and creative beyond belief happened here as a result of carbon and water. Once it started it never stopped and it completely remade our world. Carbon in water crawls and flies, respirates and synthesizes, colonizes, adapts, seeks and hides, gives birth, invents, worries, wonders, and sings. If that's not magic, then what is?

Furthermore, carbon-based life may be common in our universe. Both carbon and water are abundant everywhere we look, even out to distant galaxies. We have found carbon molecules, often mixed with water ice, in comets, asteroids, meteorites, and even interstellar dust. Experiments have shown that at least the preliminary chemical steps of life might be commonplace wherever carbon and water meet. Possible signs of ancient life have even been found in a meteorite from Mars, suggesting that this may have happened on more than one world. By contrast, the prospects for other kinds of life are dim—or at least uncertain—or such is *the view from here*—from this planet, so shaped by our kind of life, and also from inside our carbon-based brains as we look out through our carbon-based eyes.

Science strives for objectivity and uses controlled, blind, and repeatable experiments to remove the bias of the investigator. In the study of life, however, we *are* the phenomenon under investigation. This makes objectivity difficult or impossible. Thus we should be very humble in our conclusions. Arrogance and dogma, especially in an area where objectivity is impossible, are the antithesis of science.*

Sometimes I wonder if we are capable of thinking objectively about carbon-in-water. Are we deeply carbon biased? Who can blame us if we are? (Maybe a sulfuric slime-sloth could.) I am talking about a possible bias to our perceptions here which is definitely not "subconscious" in the normal sense of the word, because that implies that it is based on some ideas that reside somewhere in the mind. This bias may be built into the chemical structure of our every cell. "Pre-conscious" might be a better word. Our doors of perception are made of carbon-

---

*Such arrogance certainly will not win science many fans. Scientists often complain about the proliferation of pseudoscience, mysticism, and superstition and decry the lack of faith in the scientific method. If we added more humility and appreciation of the unknown and unknowable to our scientific education, it could go a long way toward helping us compete in the worldview marketplace.

in-water. We see the universe through carbon-tinted glasses. Our very mechanisms of thinking and seeing are constructed of these same materials. Our external and internal worlds have been shaped by four billion years of carbon evolution. How much does this limit, or define, what we see of the universe? Is our love of carbon (and our propensity to view the universe's potential fertility wholly in terms of carbon-based life) merely a bias based on our local experience, or is carbon really the only way to live? I have no idea. It's possible that this is inherently difficult for us to know. We have here an example of one world with life. That is not much to base a solid scientific conclusion on.

If life on Earth had started out using a different chemical system, would it by now have evolved to the point where it works so well that it *seems* to us like the only one possible? "Geophysiology" suggests that biological evolution always reworks its planet to optimize it for a growing biosphere. Maybe any sufficiently evolved chemical system will appear, to organisms evolved within it, as the only possible kind of biochemistry, at least when they first wake to consciousness and assess their situation in the cosmos.

LIFE OUTSIDE THE KEY OF C?

Although carbon thrives here, it may need very special circumstances to come alive. In a wide range of conditions and locations carbon spontaneously forms the small modular units, like amino acids, that could be the building blocks of our kind of life. In the right conditions these units can assemble themselves into the large, complex organic molecules of "life as we know it" (like proteins, which are long chains made of thousands of amino acids). But, those "right conditions" define a much narrower set of requirements. Carbon probably needs liquid water. It probably also needs a certain pH range (neither too acidic nor too basic), and some protection from the ultraviolet radiation which is ubiquitous in the universe. In most places carbon does not have "the right stuff" and does not behave in such a way as to be conducive to life. As far as we know it's only in the unique liquid water environment of the Earth that these little bits of carbon get together and make life. This is why our thoughts about searching for life on other worlds are so focused on looking for other planets with liquid water.

If we lived in another environment, would carbon seem so special to us? If we lived on Venus, we might not even know about organic chemistry. Carbon would be something that exists as an oxidized gas (CO or CO_2) at room temperature and does not form stable complex molecules. Maybe we would learn about carbon molecules in the interstellar medium, on comets, or on Earth. Maybe we would notice a pervasive green pigment on Earth that seems to involve carbon chemistry. We might call it the "unknown absorber."

The temperature range of liquid water informs our intuitive sense of what is too hot, what is too cold, and what is just right for life. The properties of complex chemicals like proteins and DNA are highly sensitive to their external environment. That is why "life as we know it" has such a narrow comfort range. Heat us up just a bit and we cook; our proteins "denature," falling apart into their constituent amino acids. Cool us down a bit too much and we freeze, our cells turning to ice crystals that shred our cell walls. The realm of water defines terrestrial life's ultimate boundaries. Is it the same in the universe at large? Does life always need a watering hole?

Or does the universe have other ways to solve this problem? Could it be that by only seriously considering carbon-based life we are blinding ourselves to much of the universe's true biological potential?

One of the lessons we are learning from our recent explorations of Venus is that nature is not likely to do exactly the same thing twice in the evolution of complex planets. Could the same be true of the evolution of complex life? An obvious objection to this line of argument is the following: we have not thought of any other ways to do it. If there are other possible biochemistries, then what are they? Make a countersuggestion! I am not persuaded by this objection. It's true that no one has devised an alternative biochemistry, but this could be a measure of our ignorance. I think it is safe to say that we would not have thought of carbon-based life either, if we hadn't had the Earth's example to examine and dissect.

We are still trying to learn how biochemistry works here on Earth. We do not know nearly enough to have figured it all out from the basic laws of physics and chemistry. It's no accident that chemistry is a metaphor for unpredictable human interactions: we don't know enough to be able to predict what the outcome of all possible chemical combinations will be, although we are getting better at it in some cases.

If we ever did perfect this art (or science) there would be no need for experimental chemistry.

We are still clueless about many details of the workings of our own bodies, but we know carbon chemistry fairly well because we are naturally curious about it, and there are huge economic incentives to study it. What other element can claim to be the basis of several academic disciplines (organic chemistry, biochemistry) and numerous multinational corporations (Sandoz, Genentech)? Who knows what potential, what lurking complexity, what surprises we would find if we turned the macroscope of our intellect with such sustained and focused intensity on any other corner of the periodic table?

These days many people are trying to save Earth's rain forests from destruction. This is important not only because of their role as a contributor to the global atmospheric balance and a habitat for life, and not just because of their sheer priceless beauty, but also because of the vast and largely untapped *chemical creativity* in the forest species. We are still finding substances with significant pharmaceutical potential that we haven't the foggiest how to invent or manufacture. Time and again, even in the familiar environs of Earth, we find that "dumb" nature is so much smarter and more resourceful than we are. How can we put narrow limits on its creativity elsewhere in the universe? Nature always surprises us. Why shouldn't this be true on the level of planetary environments and chemistry for life? If the universe wants to be alive badly enough, it will find a way. Nature will find the hidden pockets of order, nurture them, and shape them into biospheres which grow into and with their worlds, transforming them into magic kingdoms like, but unlike, our own. To assume that carbon-in-water life is the only kind possible could be like our historical assumption that Earth was the only world.

LOOKING FOR LIFE UNDER THE STREETLIGHT

They are ill discoverers that think there is no land, when they can see nothing but sea.
—FRANCIS BACON, *Advancement of Learning*

At a conference I attended recently, a colleague began his talk on exobiology by stating, "We assume that life requires liquid water because otherwise the problem is completely unconstrained." He went on to give a brilliant talk on the prospects for finding habitable planets.

Using carbon-in-water life as a frame for our research efforts, we have done some very good science. We determine the effects of the distances of planets from their stars and define a "habitable zone" within which life could flourish. Then we estimate the number of planets in our galaxy that may reside within these zones and could support life. There is no way to test these ideas until we find some more Earthlike planets, but this work helps us see Earth in a wider context, and it is useful for gauging the prospects of water-based life in systems of planets around various types of stars. NASA's plans for future initiatives to find other life in the universe also focus on searching for Earthlike planets and atmospheres.

If you lose your car keys at night on a long, dark street, where should you look for them? There is a joke about looking under the streetlight because they would be easier to spot. This may seem like a silly strategy, but actually it makes a lot of sense to start searching there. Since they could be anywhere, why not look first in the stretch that is well illuminated? You may get lucky and find them there. But you should not have the misconception that they are actually more likely to be under the light. And if you don't soon find them, you will need to venture out into the darkness.

If we want to search for life on other planets, then looking for liquid water is a very practical starting point. We need "well-posed" questions to study, something concrete to work on. We can't (yet) really do science to address general questions, such as "What is life?" "What forms might it take?" "What kinds of environments does it require?" and "How would we recognize it?" We need specific theories to test so we can make predictions, do experiments, and evaluate the results. So we focus our research efforts on more specific questions, like "Is there carbon-based life, similar to ours, on other planets?" and "How common might such planets be?" These questions are well illuminated by our knowledge of terrestrial biology. By investigating them we can use the power of science to chip away at the larger questions. We should certainly continue along these lines, but we should also remember that our search criteria are limited by pragmatic concerns, not by real knowledge of life's limits. This limitation winds up contributing to an air of confident consensus about the best places for life to evolve. We are looking under the streetlight with these efforts, but we usually don't present them this way.

The question of life elsewhere is a huge domain of unknown territory, a long, darkened street. We poke around the small areas in the light, which we can investigate with well-posed scientific questions while we await a more general theory of life, one that will provide more general search criteria and allow a scientific search for life in the darkened corners of the universe. Yet we are impatient to start our search, because we want to know if we are really alone. There is nothing wrong with beginning to look, as long as we acknowledge that we don't completely know what we are looking for. But should we stop at the edge of the light? Whether or not our search for nearby water-borne life is successful, eventually we may want to venture out into the darkness of more unfamiliar chemistry and relatively unexplored thermodynamic spaces.

How would we go about this? We cannot investigate in detail every possible kind of chemistry in every conceivable environment. Such a research program would be hard to fund, unlikely to yield anything useful in any given day or year, and possibly quite boring. But nature *is* conducting these experiments somewhere and has been doing so all over the universe for a long time. Who knows what's been cooked up?

Is there a less geocentric, less parochial set of criteria for detecting extraterrestrial life? Unfortunately, we don't have anything like this, at least not yet. In the meantime, another approach might be to look for unexplained complex phenomena instead of for very specific chemical signs. Perhaps we should take a more general approach, being on the lookout for signs of the unusual and the seemingly improbable. In my view, we should regard any mysterious phenomenon of global scale, especially those that involve unusual states of chemical disequilibrium or equilibrium, as *possible* signs of life. We might also look for globally pervasive unexplained absorption signatures, which could be photosynthetic pigments of some kind. Since we have no precise criteria for this quest, we must cast a wide net, or we could miss something.* With

---

*The so-called "face on Mars" definitely does not fit these criteria, mostly because it is not mysterious. It is a rock formation that looks facelike only under certain lighting conditions and has been used for commercial exploitation of the gullible. Another crater has a pretty good "happy face" in it. Does this mean that there was once a thriving "rave" culture on Mars? Be suspicious of all claims of extraterrestrial life detection.

a great degree of skepticism, but also a healthy appreciation of our grand ignorance, we should regard these kinds of features as unlikely, but possible, biological phenomena, until we find other satisfactory explanations.

LIFE ON VENUS: AN AGNOSTIC VIEW

Every age of science is so sure of itself, sure that it has almost all of the answers. In hindsight, centuries later, we often find we were far off the mark. We cannot say which conceits of the science of our day will seem hubristic and narrow to those of later ages. If I had to hazard a guess (and this is hazardous!), I would name two: One is the idea that we have almost figured out the whole universe and need only one last equation, a "final theory" that will give us all there is to know. The second is the idea that all life must be made of the same stuff we are made of.

If we relax this criterion, then the ban against life on Venus is less certain. Everyone seems to be convinced that it is completely obvious there is no life there, but several new developments have made me rethink this—some new ideas about Venus and some new ideas about life. Venus may have some of the general properties of a "living planet": a geologically active surface and a self-refreshing atmospheric and cloud system. Some scientists working on general theories about biological evolution believe there may be a tendency in the universe for self-organization, complexity and—just possibly—biology. This may or may not have specific chemical requirements.

Admittedly, this is a long shot, but I consider the idea of life on Venus to still be an open question. In any case, thinking about it is a worthwhile exercise that may help us to better understand the real limits of life.

This solar system is full of surprises. We keep finding familiar processes occurring in the strangest of places. Nobody expected to find "rivers" on the plains of Venus or "snow" on its high peaks. On Venus, sulfuric acid plays the role of forming clouds, just as carbon dioxide does on Mars and methane does on Titan. Other substances are condensing out on the mountaintops and sustaining long-distance flows. These are all roles played by water on Earth.

Planetary evolution has certain recurring themes: phenomena and structures that are common on planets with widely different conditions. When the material that plays this role on Earth is unavailable or inappropriate, then something else takes its place. Different physical conditions allow different substances to step up to the plate and perform the same function. Could it be so with life?

In this view, the statement "Life can't exist on Venus because it's too hot" may be analogous to "It can't snow on Venus because it's too hot." If rivers can form on Venus without a drop of water, maybe some kind of cells and organisms could evolve, using stuff that we are not predisposed to think of in such a role. A living cell is much more complex than a river. Does that make it more or less likely that we would recognize or imagine the correct analogy?

FISH TALES

When I was little, one of my favorite books was *McElligot's Pool* by Dr. Seuss. I used to make my mother read it to me over and over.* The book concerns a young dreamer named Marco who is ridiculed for fishing in McElligot's tiny, stagnant pool, a place where everyone knows he will never catch anything but old cans and boots. Marco answers his critics with an elaborate and imaginative fantasy. He thinks: "This MIGHT be a pool, like I've read of in books, *connected to one of those underground brooks!*" He imagines that this stream is connected to a subterranean river, and: "This might be a river, now mightn't it be, connecting McElligot's pool with the sea! Then maybe some fish might be swimming toward me! (If such a thing *could* be, They certainly *would* be!)" The rest of the book catalogs an incredible menagerie of exotic fishes out of Marco's (Dr. Seuss's) bountiful imagination, all of which *might* be parading towards his little pool to bite his hook at any moment.

Sometimes the logic we use to discuss extraterrestrial life reminds me of McElligot's Pool. Take the search for life on Mars:

Mars today is not a likely home for life as we know it. It is too cold and dry, and the surface is bathed in deadly ultraviolet radiation. And Mars is in many ways a dead world. There is little active geology, and little fluxing of energy and matter between reservoirs in the surface,

---

*I mentioned this to her recently, probably close to 30 years after the last of these readings, and she flawlessly recited some lines.

atmosphere and interior.* These are the flows that on Earth establish gradients and niches in which life can build itself.

Yet two amazing discoveries we've made about Mars have gotten us excited about the prospects for past (and just possibly current) life there. First, orbital photos showed that there has almost surely been running water, and probably standing water there in the past. This implies that Mars once had a warmer climate, which probably required a thicker atmosphere that would have protected the surface from deadly ultraviolet rays. In its early history, Mars may have been more like Earth, a place where carbon molecules could have found many a friendly watering hole to evolve into life. In 1996, studies of a meteorite which blasted off of Mars in a large impact 15 million years ago and landed in Antarctica 13,000 years ago yielded the second discovery: This rock contains possible microfossils and chemical clues which could be the signature of ancient carbon-based life on Mars.

Even though the environment has changed over the eons, become cold and dry, iced-over and irradiated, maybe life—if it got started there—has hung on somehow. This possibility prompted the Viking biology experiments we landed in 1976. We found some unusual chemistry in the soil of Mars, but we didn't find life. Viking also showed us that there are basically no organic molecules in the soil of Mars, as might be expected in an ultraviolet killing field.

These elaborate instruments tested the hypothesis that the surface of Mars is covered with Earth-style life. The results seem to be negative, but that has not stopped our search for life there. If you are attached to the idea of life on Mars, you can retreat to the positions that: 1) Life might be isolated only in certain pockets of the surface, so any random landing site would be sterile; 2) There is life underground, just below the depth to which the landers sampled; 3) Mars might still have pockets of warm water, in hot springs, supporting colonies of life; 4) Life there is chemically very different and would not respond the way the designers of the Viking biology experiments expected; 5) Life may have evolved on a past Mars with a more Earth-

---

*There are active seasonal cycles of condensing and sublimating CO_2 and H_2O, but these don't involve much chemistry.

like environment but died out when the climate changed, leaving fossils that we can find.

The naysayers might declare "You'll never find life on that scrawny little planet, with its deadly ultraviolet radiation, freezing temperatures, lack of organic chemicals, and poisonous atmosphere." We dreamers, though, see some dried up river beds and think: Just maybe these were carved by running water which *might* have been connected to a mighty ocean in the past. And that ocean just *might* have had organic molecules that behaved like Earth's, and just maybe a thriving biosphere evolved. And *maybe* even this life is still there hidden in underground caves or hot springs. Or, these signs of past water *might* lead us to fossils of past life. Faced with the logic of McElligot's pool, there is no way to disprove the existence of life on Mars. Anyone who wants to believe is free to do so, in no danger of violating any scientific principles.

I do not mean to ridicule the dreamers. I am firmly on the side of Marco and the believers in possible life on Mars. The argument for the possibility of fossils is especially compelling. Both Mars and Venus could have had similar environments to Earth when life was getting started here. It's possible that the same magic started up in the oceans of all three worlds. Mars, because it is geologically dead, has preserved its past (including any fossils) best. If this fish story seems unlikely, it is also worth considering how important and exciting success in such a search would be. And the effort is inherently worthwhile. Fishing is a good way to spend time even if you don't catch anything. The search for life, futile or not, can't help but teach us a lot about Mars and terrestrial planets in general. So, why not look?

Mars could have isolated pockets of hydrothermal systems, supporting underground bacterial colonies. But two ideas about life as a planetary phenomenon argue against this. If Vernadsky is right, life is a widespread property of a planet that covers it entirely if it exists at all. The signs of life will not be subtle; living planets will exuberantly announce themselves to the most casual observers. The terrestrial example, where life has tenaciously and aggressively expanded and colonized, finding toeholds almost anywhere you could imagine, certainly supports this view. If the "gaians" are right, then life should be obvious from a distance due to its effects on the atmosphere. These arguments seem plausible, but they are not completely compelling because we don't really know what life will do to a planet and what we should look for.

I've presented the Mars story in a way that draws attention to the importance we place on water. If our search continues to yield negative results, we will have to invent new possible hiding places for Martian bugs. But we cling to the signs of past water on Mars as a possible link to our own biochemistry.

If we relax, for a moment, our rule that life must cling to water, then the prospects for life among the terrestrial planets seem a bit different. As discussed earlier, Mars is the most Earthlike of the other planets in its *surface conditions*, but Venus may be the most Earthlike in its *activity*. All three planets started out young and restless, with warm, churning interiors and water flowing on their surfaces. But they've gone their separate ways. Mars cooled off, ceased its geological activity, and lost most of its atmosphere to space. Its water either left for space as well or else is hidden, probably frozen, underground. Mars today is a dead world, still wearing the surface scars of the solar system's violent birth.*

Venus, too, lost its ocean, but it retains an active, churning interior and a surface that has been reworked many times by processes that are apparently still ongoing.† It has a nonequilibrium, chemically restless atmosphere and supports global clouds that function in some ways like Earth's oceans.

Again we must ask what life really wants and needs. If life follows water, then Earth may be the only living world in our solar system. Life could have started on all three worlds, and been banished from Venus and Mars as they lost their oceans. It could also conceivably be clinging to isolated watering holes beneath the surface of Mars.

If life needs an active planet, with continuous sources of energy and nutrients, and is less picky about its chemical choices than we have imagined, then Venus may be the best hope for other life nearby.

---

*By the way, if Mars is uninhabited, that makes it a *better* place to colonize. We don't have to worry about disturbing the natives or catching any new germs.

†As I've discussed, we don't know if Venus was born as wet as Earth, but it surely had some water to begin with.

READER'S ADVISORY: THIS BOX CONTAINS
EXPLICIT SPECULATION!

LIFE SIGNS?

I have now crawled so far out on a limb that I see no reason to try climbing back. I may as well jump! So let me propose some possible signs of life on Venus. Let me state clearly that **I regard each of these possibilities as extremely unlikely.** But in the search for life the criteria for declaring a negative result are different than in, say, the search for volcanoes, since we don't really know what we are looking for.

Here are four phenomena that could be signs of life on Venus:

1) *The atmospheric "superrotation" could be created by life.*

This is one of the most obvious and large scale unexplained features of the planet. The dark ultraviolet markings whip around Venus in swift and colossal winds, circling the planet every four days. From the point of view of any Venusian bugs that want to use sunlight for energy, the supperrotation would be a major plus because the night is so long there. The planet may rotate too slowly for photosynthesis unless you have something like a super-rotation.

Perhaps Venusian organisms transform solar energy into mechanical energy to drive the winds. Radiation in the clouds might be absorbed by microbes suspended in the cloud particles, and the resulting temperature differences could cause biologically controlled winds to blow. In fact this mechanism, called "radiative dynamic feedback," has been invoked in less radical, non-biological form, to explain some of the observed structures in the clouds.

Picture the clouds as a kind of super-organism that absorbs light in such a way to keep itself spinning so no part of it has to stay in the dark too long.

2) *Maybe the "unknown ultraviolet absorber" is a photosynthetic pigment.*

We love sunlight, but not all of it. We like the visual portion, which illuminates the surface of our world and fuels the life upon it. Shorter wavelength, ultraviolet light is lethal to us, because this higher energy radiation rips apart carbon bonds. Imagine creatures that have evolved some alternative chemistry to tap the huge amount of energy in the ultraviolet portion of the sun's spectrum. Maybe they think that no life could possibly be sustained by the lackluster, low-energy photons of visible radiation. Photosynthetic life on Earth makes use of chlorophyll, the ubiquitous green pigment. If Venusian life has evolved to take advantage of UV light, this might be done in the form of a pigment that absorbs ultraviolet. If this is some complex chemical unknown to us, that could explain why we have had such a tough time figuring out the identity of the "unknown ultraviolet absorber."

3) *Maybe Mode 3 cloud particles are alive.*

As I described in Chapter 3, Mode 3 are the odd, large cloud particles. There have been a lot of contradictory measurements of their size, shape, and composition, and we do not yet have a good description of them or explanation for them. They may not be spherical in shape: they may be solid crystals. Some measurements suggest that they are made of sulfuric acid, like modes 1 and 2, but there is also evidence of more exotic chemicals, like chlorine or nitrogen compounds. They might be sulfuric acid droplets surrounding some other substance, perhaps one that forms a solid core suspended in the interior. Could those strange cores be some kind of creatures?

An exobiologist friend of mine dismisses the idea of cloud-life anywhere. He asks, if this is possible, then why aren't the clouds green on Earth? It's a good point. Life on Earth is opportunistic, inhabiting every niche it possibly can. If cloud-life were possible, Earth's clouds would presumably be full of it. This argument once again limits us to extrapolating from the known, which is generally a safe tactic in science. I am not convinced it is a good idea when thinking about exobiology.

After all, the clouds of Venus are a very different kind of place from the clouds of Earth. They are a much larger, more continuous, more stable environment. The difference in stability is important. Models indicate that cloud particles there last for several months, much longer than those on Earth. An enterprising species could easily fit its life cycle into those months, seeding other cloud particles with its progeny before falling into the hot winds below.

The clouds are where most solar energy is absorbed and the site of some of the greatest chemical and energetic flows on Venus.

They are also full of interacting liquid droplets, which may have been important in the origin of life on Earth. And maybe in their own way, the clouds of Venus are "green." Green for Venusian life might be in the near-ultraviolet. This may be pushing things, but if I were devolved matter thinking of establishing a biological beachhead on Venus, I would take a good look at the clouds, the ocean of Venus.

It's also worth noting that the temperature up there is much milder than at the surface. In fact, conditions in the clouds of Venus are not too different from those at the surface of the Earth. There is a level in the clouds (about 33 miles up), where the atmospheric pressure is about 70% of the pressure at sea level on Earth, and the temperature is a balmy 107 degrees Fahrenheit. For ballooning at this altitude on Venus, you would need only a thin, acid resistant suit, an oxygen tank, and a large supply of cold lemonade. It's cool enough for liquid water, and small amounts of it exist there (in a strong sulfuric acid solution).

Still, something in my gut tells me that the clouds of Venus are not a good biological habitat. That something is stomach acid, hydrochloric acid. Acid eats organic molecules. The Venusian clouds could be a niche for "life as we know it," depending on carbon in water, if it weren't for all that damn acid. But we have recently discovered that there are thriving colonies of bacteria living in some people's stomachs, much to their discomfort (the people, not the bacteria). This is one example of the innumerable ways that we seem to be still learning, all the time, of surprising new niches where life can live on Earth. The crafty bugs in our guts have found a way to build an insulating lining protecting their innards from the acid environment. Who can say that Venusian bugs couldn't do the same? Earth life can adapt to a highly acidic environment. Still, the Venusian clouds are much more acidic than your stomach, unless you are having a REALLY bad day, but some bacteria on Earth thrive even in concentrated sulfuric acid.

4) *The highly reflective mountain tops could be covered with life.*

Some kind of transformation happens to the ground all around the planet above an altitude of 13,000 feet (which corresponds to a temperature of 820 degrees). I have mentioned that this could be analogous to a snow-line. What about a "tree line"? I am not suggesting that the high peaks of Venus are forested with Douglas Firs, but maybe this temperature boundary represents the edge of the "habitable zone" on Venus. Perhaps some non-carbon-based equivalent to lichen grows on Venusian rocks below a certain temperature, feeding off of the disequilibrium sulfur gases. These creatures could be causing some chemical change in the ground which concentrates radar reflective materials.

IS VENUS COMMONPLACE?

Now that I have gotten those flights of fantasy out of my system, it's time to come back down to Earth—or at least to Venus. Here is another reason why the possibilities of life on Venus may not be completely academic to students of exobiology: there are probably more Venuslike than Earthlike planets in the universe. Admittedly, our available sample of terrestrial planets is so limited that any statement about what planets in general are like should be taken with a huge grain of salt. When you hear anyone make definitive statements about planets in the universe, your b.s. detector should go immediately into the red. Nonetheless, we can observe that in our ensemble of terrestrial planets, to the extent that there is a "normal" atmosphere, it is mostly carbon dioxide, with traces of other things. Earth is the misfit, weirdo planet, and we are not sure why. We know that life is involved, but cause and effect here is muddled. Recently, we have decided that the best explanation for the existence of our strange, giant Moon is that Earth got clobbered by a Mars-sized stray planetesimal very early in its history. This event may have had major effects on the physical characteristics of the Earth, and in particular may have drastically changed the atmosphere in ways that no one has figured out yet. Do we owe our weird atmospheric evolution, and subsequent friendly locale for "life as we know it," to this fluke? If so, then Earthlike places might be very rare. Most Earth-sized planets might be more like Venus.

ARE WE ALL VENUSIANS?

It is just possible that our voyages to Venus are a journey back to our oldest ancestral home. If Venus, Earth, and Mars all started out with oceans full of organic chemicals, this "soup" might have cooked up carbon-based life on any of them. If life formed anywhere on the early terrestrial planets (and it seems that it did . . .), it could have spread to all of them. In their early years, the planets may have been highly contagious. Their earliest environments were often disturbed by large impacts from the heavy bombardment that marked the final stage of planet growth. Such events can launch rocks off of one planet that eventually land on another. We know that this happens because we have already found about a dozen meteorites from Mars here on Earth.* Early life could have traveled from planet to planet, hitch-hiking across space on the impact debris that was blasting between them. In this scenario our carbon-based ancestors could have originated on Mars or Venus—or Earth could have infected these other worlds.

Venus might have been the best place of all for organic life to get started. Our Sun has been heating up steadily over the 4.5 billion years of its existence. Sunlight was dimmer by about 40 percent when the Sun and the planets were very young. This means that, depending on the details of atmospheric evolution, all of the terrestrial planets might have been much colder. Earth and Mars may have been too cold, and Venus "just right" for life of the water-borne variety. If life did form on Venus and spread to the other planets, it must have emigrated quite early, since our ancestors were certainly here 3.8 billion years ago. On Mars the poor little buggers would eventually have been freeze-dried by subsequent global change, whereas on Venus they would have been broiled, unless they found a way to live in the cool clouds and eat acid.

We may have gotten started on Venus. If so, then our spacecraft are the first calling cards of a prodigal sister, returning home after living for billions of years on another world.

LONG LIVE MAGELLAN

We know for sure that there are signs of life at a few places on Venus. We know because we left them there—the smashed, corroding remains

---

*One of those Martian meteorites, Nakhla, actually struck and killed a most unlucky dog when it fell in Egypt in 1911. Another, ALH84001, contains the now-famous putative microfossils. There are probably Venusian meteorites, too, but we haven't identified any yet.

of our inquisitive machines. The most recent addition to this smattering of Earth junk is whatever is left of *Magellan*. If *Magellan* could have asked us, it might have requested to be cremated and have its ashes scattered around Venus. That's pretty much what happened to it. There could be no more fitting end for the spacecraft that brought a world alive for us.

By early October 1994, it was clear that the end was at hand. We had completed most mission goals, and funds were running dry. The spacecraft was losing power rapidly, its solar panels finally giving out after seeing *Magellan* through more than fifteen thousand orbits. *Magellan* could have been left circling around Venus, but its low orbit would have decayed eventually, and it would have met the same fate. Instead, it was done in by an "assisted suicide" that allowed us to perform one final experiment. On October 11, 1994, five and a half years after it was launched from Earth, we fired thrusters to lower *Magellan's* orbit into the upper atmosphere. The winglike solar panels were rotated to make a windmill shape, and we measured the increasing thickness of the upper atmosphere by watching how *Magellan* spun as it plunged. *Magellan* died as a kamikaze corkscrew for science, sending home new information until the end.

Our tracking stations lost *Magellan's* signal for the last time at three in the morning, California time, on October 12. We can pretty much guess what happened after that, as *Magellan* finally entered* the world it had circled and scrutinized for years. The solar panels and large main antenna must have been torn off first. Then various smaller pieces would have started to rip off, some igniting in the growing heat from friction with the increasingly dense atmosphere. But *Magellan* entered fairly slowly and probably did not burn up entirely. The larger chunks probably fell to the surface, where they would have landed gently because of the dense atmosphere, like shells settling to the bottom of the sea. Judging from *Magellan's* location at the time of its final transmission, some pieces may have landed on the high peaks of Maxwell Montes. Maybe someday the Smithsonian will offer a prize for the first *Magellan* part found and returned to Earth.

---

*This has been called *Magellan's* atmospheric "reentry," but that doesn't seem quite the right term for a spacecraft launched from another world.

We have not yet thoroughly analyzed, much less understood, the huge wealth of data dumped on us by *Magellan*. It takes a while to digest a world's worth of images, reevaluate our home world in light of the data deluge, and develop a new understanding of Earthlike planets. The big picture of Venusian evolution is still shrouded in clouds, now patchy and thinning in places, which will continue to be slowly dispersed by mature reflection.

We've learned a lot about Venus and we will learn a lot more by carefully studying our *Magellan* images over the decades to come. But we still need a lot of answers that we will have to go back to Venus and ask for.

BACK TO VENUS

How soon and in what style will we go back to Venus? The answer largely depends on political and economic developments, which are even more complex and harder to predict than planetary evolution. There is no shortage of plans and ideas for further exploration, some of which would not be too expensive. The more we learn about other planets, the cheaper it gets to learn more. We don't need another huge mission like *Pioneer* to learn an awful lot. Now we know better what questions to ask, and how to rephrase the questions more precisely so that Venus is less likely to give elliptical answers. Knowing what we know now, we could send some small, low-cost, focused missions that would be very fruitful.

We could put a small craft with a few carefully chosen instruments in orbit around Venus, or land a very small package on the surface, for a ballpark figure of $200 million, less than a buck from each American.* More ambitious missions involving multiple landers or experiments requiring long survival times at the surface would be considerably more expensive. We might pursue these as international collaborations, sharing around the Earth the cost of exploring Venus.

You may have noticed a pattern in my descriptions of our planetary missions. There is always a grandiose list of questions we hope to

---

*This is how much it cost to make the movie *Waterworld*, and is roughly the sum being sought to build another football stadium in Denver. (I told you it was a ballpark figure.)

answer with the promised new observations. Later, when I describe what we have actually learned from the *Mariners, Veneras, Vegas, Pioneers,* and *Magellan,* there is a lot of ambiguity, and the results must be couched in caveats and uncertain assumptions. We hardly ever get complete answers to the original questions, but we always discover things we were never looking for.

This is just a reflection of the way our science works. To justify a mission (to ourselves, to Congress, to the public), we need a list of well-defined questions, but real planets are complex, messy, and full of ambiguity. Many of the benefits of exploration simply can't be anticipated. If we knew exactly what we would find there, why go?

Having said that, here are some questions for future missions:

What is the unknown ultraviolet absorber?

What drives the superrotation of the upper atmosphere?

What is stabilizing the climate?

What is the chemical composition at the very bottom of the atmosphere?
Our evidence about the part of the atmosphere actually in contact with surface rocks is vague, contradictory, and confusing.

What is the composition of the "noble gases"?
Pioneer and the *Veneras* gave us only vague answers. Precise values could help us pin down the history of the atmosphere.

What is the effect of a "supercritical" lower atmosphere?
This is something I haven't discussed here, and no one has worried about it too much yet, because we have had so much else to think about. The atmosphere at the surface of Venus is neither a gas nor a liquid. At the high pressures found there, that distinction breaks down, and CO_2 is "supercritical." We are used to thinking of the atmosphere as a gas and that Venus has no liquid oceans. In fact, the lower atmosphere of Venus may truly be as much a liquid as it is a gas. This could have a major influence on the chemistry and radiation down there, and other aspects of the environment as well. How would this affect the climate-sensitive chemical feedbacks between "air" and rock at the surface?

How old are the various kinds of terrain?
For example, how old are the tesserae? Are they all the same age? How old are the plains? What are the ages of flows around young shield volcanoes?

What types of minerals and rocks are present?
Ideally, we would want answers from many locations, sampling rocks from tesserae, from the rolling plains, from a young shield volcano, from a pancake dome, and from Ishtar and Aphrodite, for starters.

What is the "snow" at high altitudes?

What is the present rate of volcanic eruptions?

What gases are volcanoes exhaling now?

What is the interior structure?
It would really help us to know the thickness of the lithosphere and how much it varies from place to place, and the overall density distribution and motions of the mantle. Does Venus, like Earth, have a solid inner and liquid outer metallic core?

What does the surface look like up close?
Magellan's radar uncovered the surface for us at a certain scale, but we still don't know what it would look like to us, up close, in visible light. The answer will be different at different places, and we won't know until we send cameras.

Here are several possible new missions that could answer some of these questions (and probably force us to ask many new ones that we can't even imagine now):

ORBITER: Since the last time we had an orbiting spacecraft to study the atmosphere of Venus, we have discovered the near-infrared windows discussed in Chapter 3. An orbiting spacecraft with cameras tuned to these windows could do fantastic things. This instrument would be similar to the NIMS instrument on *Galileo* (described in Chapter 3), but optimized for Venus. Even in *Galileo's* brief fly-by of Venus in 1990, NIMS provided us with an invaluable three-dimensional snapshot of the clouds. Long-term monitoring with such an instrument would reveal changing structures and motions at all levels in the clouds and possibly provide clues to the nature of the superrotation. It would also allow us to peer through the infrared windows and observe changes in the lower atmosphere. Maybe we would find the "smoking volcano"—catch Venusian volcanoes in the act of erupting by sensing the infrared signatures of their gas plumes. Thus, even from orbit, we could learn about the composition of the volcanic gases and the ongoing eruption rate. An orbiter could also carry an ultraviolet spectrometer for further study of the unknown absorber.

AIRPLANES AND BALLOONS: The Soviet *Vega* mission of 1985 pioneered the use of balloon stations in the Venusian atmosphere. We could do a lot more along these lines. By setting balloons adrift in the atmosphere, we could learn about circulation patterns. Onboard instruments could measure atmospheric composition and study cloud drops. At lower altitudes, cameras on balloons could photograph the surface as they circled the planet, getting a free ride from the superrotation.

A slightly more elaborate approach to the balloon exploration of Venus is the *aerobot*. These are robotic balloons that could adjust their buoyancy to control their altitude.* They could make brief forays down to the hot lower atmosphere and possibly even land on the surface to do experiments or drop off small instrument packages. When they begin to get too hot, they could retreat to higher altitudes to cool off, ride the global winds around the planet, and radio home their findings from the latest venture to the depths.

Drone airplanes or blimps could fly low over the surface, photographing large areas.

ENTRY PROBES: One probe, building on the heritage of the *Pioneer* designs but carrying a new generation of experiments, would answer many of our lingering questions about the atmosphere. Experiments pinning down the composition of the lowest part of the atmosphere and getting more accurate measurements of the noble gases would help us understand the origin and history of the atmosphere

---

*We can do this with a "reversible fluid" that changes back and forth from gas to liquid. When you cool it and it condenses to a liquid, the balloon is denser than the surrounding air and descends. Heated, the fluid evaporates and becomes a gas, lowering the balloon's density and causing it to rise.

and the current climate balance. Such a probe could also study cloud particles on the way down. Another idea is a "multiprobe mission" that would deliver a large number of very simple probes consisting of radio transmitters. By tracking their positions and motions as they fell, we could gain a comprehensive picture of the global atmospheric circulation.

LANDERS: We would love to land more instruments to study the chemistry of surface rocks, test the air, and take more pictures. Now, after *Magellan*, we know where to look. A sample each of tessera, young flows from a shield volcano, plains basalts, "river" beds, pancake domes, and shiny highlands would clear some things up.

SAMPLE RETURN: As sophisticated as some of our robot spacecraft experiments are, they don't hold a candle to what human scientists can do in a good lab on Earth. You can do only so much remotely with a robot lander. To study Venus rocks in detail, we would need to send a probe that could pick some up and bring them home to Earth. That is the only way to measure the age of a rock from its radioactive decay products. Ideally, we want samples from several different places in varied geological settings to get a handle on the age distribution. (This takes us well out of the $200 million range, by the way.)

Here's a sick thought: Some recent research suggests that Venus is a common destination for fragments blasted off our Moon in large impacts, so Moon rocks are probably scattered around its surface. When we do finally get a mission to go to Venus and bring back a rock, we had better be careful not to grab one of these expatriate Moon rocks. That would really screw up our analysis!

LONG-DURATION SURFACE STATIONS: Studying Venus is like opening one of those Russian dolls with another doll inside and then another inside that one. Each layer opens to reveal another layer of mystery. We've finally found our way through the clouds to see the surface, but to fully crack the puzzle of Venus we will need to look below the surface. A network of seismic stations could allow us to do that. If we could map the interior density structure of Venus as we have mapped Earth's, much of our uncertainty about how surface structures manifest the motions below would instantly be resolved. Seismometers would allow us to see how another Earthlike (and Earth-*sized*) world has organized its insides.

This would not be cheap or easy. Any experiment requiring long-term operation on the surface would require a whole new generation of electronics and materials that could survive and function at Venus temperatures. But we could do it if we set our minds (and our wallets) to it. We are now studying the idea of a mission that would place three seismic probes on the surface, with refrigeration to keep them cool long enough to gather some data and return them home. Of course, we don't know how seismically active Venus is, so any experiment like this without long expected survival times is risky.

Of all these ideas, the two that most excite me are cameras and seismometers. A seismic network anytime soon may be prohibitively expensive, but we *must* send cameras to photograph the surface of Venus. There are surely places of fantastic beauty and complexity with intricate details and colors invisible to orbital radar. A low-altitude flying machine that covers a lot of ground would be the ideal platform. First, just to guarantee that we find something interesting, I'd go to some of the young volcanic areas. If you look up close, young volcanic places like Hawaii show a riot of interesting and diverse detail (see Figure 6.2).

On Venus you should see just as many interesting forms, but they would be different because other kinds of lavas are erupting into a very different environment. On Earth, undersea volcanoes produce bizarre pillars and pillows of lava unlike anything seen on land, and you could bet that the volcanoes of Venus would be full of delightful surprises as well. Also, since erosion on Venus is so incredibly slow, many geological forms would be remarkably well preserved. This is a characteristic of some of the most beautiful places on Earth. We could fly over the high peaks of Venus and see whether they look shiny in reflected light, then head down an ancient winding river valley. We could cruise low over Devana Chasma and see how closely it resembles the East African Rift Valley. If we are lucky, we could catch a young volcano in the act of creating brand-new parts of Venus.

I have to admit, my desire for photographs of Venus is aesthetic as much as scientific: I want to know what it looks like. But why even make the distinction? Such an endeavor could not help but be scientifically useful. To see is to gain understanding. The quest for vision and the quest for scientific knowledge blend seamlessly. Beauty and mystery are reason enough to explore.

6.2 **Fresh Volcanic Flows:** *Volcanic flows on Earth display a wide range of shapes and textures depending on the type of lava and the eruption conditions. These pictures show (top) a recent flow of ropy lava (John Spencer) and (bottom) "toes" of fresh lava (Mary Urquhart), both in Hawaii. What will Venusian flows look like up close?*

How much money should we spend on future Venus missions or planetary exploration in general? It's not easy to say, at a time when we have many urgent needs here. Our species is starving and we are trashing our planet. Our schools are falling apart, funding for the arts has all but dried up, and our inner cities are so troubled. Tough times require tough choices. But we are learning how to explore more cheaply all the time. I believe that maintaining a robust program of robotic exploration is a good deal in the long run. To gain some perspective, consider that one B-2 "stealth" bomber costs about a billion dollars. For less than the cost of a fleet of these, we could fly *all* the missions listed above. The benefits are as intangible as they are ultimately vital to maintaining and enhancing the planetary perspective that may save us from ourselves.

A LONG HOT VENUSIAN AFTERNOON

If people ever do go to Venus, what kind of a place will they find? What's it really like there? As on Earth, it depends on where you go. But there are some qualities of the environment that would immediately strike anyone who landed there and stepped out of her ship. Let's assume you are well insulated against the heat and pressure, so you will live long enough to perceive other things. You would probably notice first the quality of the light.

We think of Venus as a bright, luminous planet, but if you went and spent some time on the surface, the light would seem dim and diffuse. It's relatively dark there for the same reason that the planet glows so brightly in the skies of Earth: most of the light is reflected back into space by the clouds. However, there is plenty of light at the surface to see (like a deeply overcast day on Earth), and your eyes would quickly adjust. The most striking characteristic of the light would be its deep red cast, a permanent sunset color that results from scattering of light by layers of thick atmosphere and clouds. Looking up into that deep red sky, you might learn to judge the time of day by the slow movement of brightness across it, but you would never see the Sun, and there would be no shadows (terrestrial vampires would feel at home).

I said earlier that the sky is an unbroken, unchanging overcast. I was repeating what I've heard many times, which always seemed reasonable enough. But now I'm not so sure. Now we know how patchy the clouds are. As described in Chapter 3, recent discoveries in the near-infrared windows showed us that the bottom layer of clouds, where most of the cloud mass is and most of the light is blocked, is full of holes. Although we never saw them with Earth-based telescopes or orbiting spacecraft not tuned to the right infrared channel, we now know that there are thin and thick patches of all shapes and sizes down there. On the nightside, in the infrared, the places where the clouds are thin appear as "hot spots" from which the heat radiation shines out to space (as seen in the *Galileo* picture in the color insert). From the surface of the dayside, I suspect we would see these as brighter patches in the sky. On overcast days here on Earth we see bright and dark patches of sky where the thickness of the clouds varies. I think we would see the same thing on Venus. The contrast between the "hot spots" and "cold spots" on the nightside is huge—some places are ten times as bright as others. Some of the spots are hundreds of miles across. I think the biggest and brightest patches could be seen from the ground. If so, the sky of Venus becomes much more interesting because it would be constantly changing.

Furthermore, the superrotation of the upper atmosphere would be rendered visible from the ground. These cloud patches would move a thousand miles from eastern to western horizon in about five hours. What a sight that might be: giant shifting shapes of light and darkness slowly forming and dissolving as they traversed the sky thirty miles overhead. You might sense subtle changes in the light around you as large thick and thin patches of cloud passed by above. As you stood on the ground, the air around you would be very still, perhaps broken occasionally by a very light gust of searing wind. But glancing upward, you would see a constant reminder of the raging winds that circle the planet.

That's the daytime. The nighttime experience could be considerably stranger. At midnight the sky is completely dark, and the main source of illumination is *the glowing of the red-hot ground*! The surface is so hot (almost 900 degrees) that its thermal radiation would produce a faint red glow. The darkness might also occasionally be broken by a faint flash of lightning from high above in the clouds.*

Not just the quality of the light but the pace of its changes would tell you immediately that you are not in Kansas anymore. On Earth, day and night last roughly twelve hours, depending on the season and your latitude. The terminator (the line between day and night)[†] sweeps across the landscape at one thousand miles per hour at the equator, and is slower at higher latitudes.[‡]

On Venus, day and night each last fifty-nine Earth days, or about two Earth months. The terminator is a diffuse boundary that strolls from west to east at a leisurely eight miles per hour at the equator. At a high latitude—say, Maxwell Montes at 70 degrees north—it would move at three miles per hour. So if you stayed in one place, you would

---

*We are still not sure whether there is lightning on Venus. We've recorded some suggestive electrical pulses, but we have yet to observe the telltale flash.

[†]Not a beefy robot sent from the future.

[‡]You multiply this by the cosine of your latitude, so that at 45 degrees it moves at about seven hundred miles per hour. It also changes with the seasons.

have to wait a very long time for the light to change, but it might be worth the wait. As twilight approached, the sky would very slowly dim and the glow of the ground would become noticeable. Maybe it would feel like those enchanted times on Earth when the landscape is illuminated by a slowly shifting balance of full moonlight and fading twilight. Dawn would bring a similar slow changing of the guard from ground glow to sky glow.

This leisurely pace of change could have its advantages. On Earth, if you are traveling in a jet aircraft, you can play tricks with the time of day. Traveling westward at the right speed, you can make sunset last for a very long time. Traveling eastward, the night passes quickly. On Venus the changing of the light is so slow that you could do these things on foot! Remember how Saint-Exupéry's Little Prince kept moving his chair around his asteroid to watch the sunset over and over again? Similarly, if you had a favorite time of day on Venus, where the light was perfectly balanced between ground and sky, you wouldn't have to wait for night to fall. You could just walk there.

Some of the most interesting places to be on Venus, in terms of the quality of the changing light, might be near the north and south poles. Here you might find a perpetual twilight, a balance between the constant ground glow and the sky glow that *slowly* travels around the horizon, as Venus spins.

Because its spin axis is not tilted like Earth's, Venus has no seasons, but the changing time of day there has the pace of seasonal change here. A monthlong afternoon is followed by several weeks of twilight before night fully falls. All of this refers to changes in illumination, not temperature. If you become overheated during a long Venusian afternoon, and you want to cool off, don't bother to wait for dark. Not only is it a long wait, but the temperature doesn't change. Going north or south to the poles won't help either. The thick atmosphere of Venus redistributes heat so that time of day and latitude have no effect on surface temperature. If you want instant heat relief, climb a mountain.* Everywhere on Venus, it gets about 5 degrees cooler for every thousand feet of altitude. The high peaks of Maxwell, towering thirty-

---

*Summertime in Tucson, where I lived for seven years during graduate school, is pretty much like Venus, always hot in the desert, day and night, always cool up in the mountains.

eight thousand feet above the plains, are a cool 710 degrees in the shade.

Walking on Venus might be easier than on any other planet, if we could design a survival suit that was not too cumbersome. Remember the images of *Apollo* astronauts stumbling and bumbling around the lunar landscape, occasionally falling and taking expensive pieces of equipment with them? They had to devise a whole new way of walking, or hopping, to get around in the Moon's low gravity, which is one-sixth of Earth's. Our way of getting around on our planet is deeply tied to its gravity. But your weight on Venus would be only 10 percent less than here, so you would not need a new way of walking. In fact, it would give you an added lift, an extra spring in your step, without really throwing you off.

As for what it's like at different locations, there are probably as many answers as there are places. Personally, I would love to stand on the banks of one of the rock rivers and see if it really looks like a river on Earth, or hike up the side of a shield volcano, searching for a bubbling lava lake. We won't really know what it's like there until we go and find out.

OH WE, LIKE CATS

But will we go? We've sent our machines there, and will continue to do so. But will humans ever walk on Venus? Human exploration of space has slowed to a crawl. This seems to be part of a larger collective lack of nerve, a wave of self-doubt and fear of the future that we are experiencing as we approach the millennium. We are wondering what we have done to ourselves, other species, and our world. We are struggling with our addiction to high technology: we need it, we love it, and yet we have not completely learned how to live with it in a non-self-destructive way.

Maybe we need to slow down, catch our breath, and get our own house in order before heading out into space again. After all, look what happened the last time. We went to the Moon without really knowing what we were doing or why we were doing it, and it didn't stick. We then retreated so thoroughly that we can't even get back there today. *Apollo* was quite an achievement, but when we send people to Mars or even eventually to Venus, it would be better to have a clear, long-term plan.

I have a fantasy of cloud cities on Venus, huge enclosed habitats suspended from giant balloons. At a certain altitude where the temperature and pressure would be comfortable for us, we would mostly just have to keep the air fresh, maybe by collecting solar energy to make oxygen from CO_2, or better yet, by growing plants to do the job for us. Why should we bother to do such a thing? I don't know. These could be research stations, or maybe there will be some economic incentive, something rare or beautiful found only on Venus. Or maybe, in the very long run after we have solved all our major problems here on Earth, we will go just for the hell of it, because it is there. (Yes, now I really am fantasizing.) (See Figure 6.3.)

Anyway, it doesn't really matter what I say here or what anyone else writes in any book. Generations from now, I think, we will explore the clouds and the surface of Venus in person. I can't think of a rational reason to go, but we humans will always keep exploring, perhaps because it is our destiny, or because we can't resist knowing what lies beyond that closed door. If you have ever lived with cats, you know that they will get into anything new or mysterious that appears in the house. Any unexplored frontier compels them. They will keep trying until they get inside. Humans are a lot like cats, so curious that we must poke our heads everywhere, look around every corner, under every curtain and behind every door, even when it might get us into trouble. We will continue to explore the universe for the same reason cats explore closets. Like cats, we also sometimes have more impulsive curiosity than foresight. It's easy and fun to climb up the tree, but now how you gonna get back down? (Where you gonna bury that nuclear waste?) But we are capable of learning. You know what curiosity did to the cat, but we are a bit smarter.* I know that's not saying much, but if our curiosity is augmented with wisdom, if we integrate what we find along the way into a new, larger, and longer perspective, I think we will go far.

TERRAFORMING AND VENERAFORMING

Earth First! We'll mine the other planets later.
—BUMPER STICKER SEEN OUTSIDE TELLURIDE, COLORADO

---

*If physically somewhat less resilient, and each possessed of eight fewer lives . . .

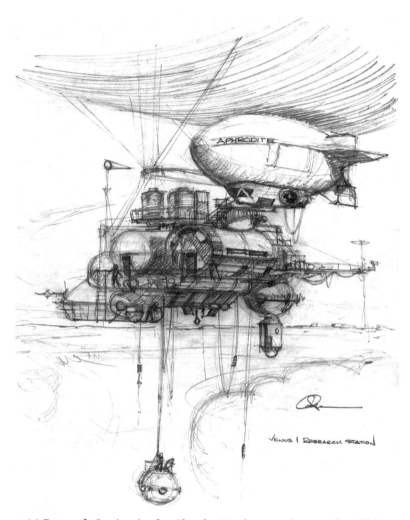

6.3 Research Station in the Clouds: *Maybe someday people will live and work in stations suspended in the clouds of Venus, powered by solar energy or wind turbines. (Carter Emmart)*

The mutation from terrestrial to interstellar life must be made, because the womb planet itself is going to blow up within a few billion years.... Planet Earth is a stepping stone on our time-trip through the galaxy. Life has to get its seed-self off the planet to survive.
—TIMOTHY LEARY, *Terra II*

Venus today may not seem like a place where anyone in her or his right mind would ever *want* to go. Could we remake it in Earth's image? This idea is as tantalizing in some ways as it is horrifying in others.

Various schemes have been dreamed up about how we might one day alter the environments of other planets to be more like Earth, so that we could go live on them without needing suits or domes. This is called "terraforming." Most schemes focus on Mars because it already has the most Earthlike surface conditions and so would need the least work.* These plans are somewhat cartoonlike: "All we have to do is add 20 zillion tons of oxygen and then plant some geraniums." They tend to focus on what is needed for life and ignore hazards that may be present. For instance, if we created a warm, oxygen atmosphere on Mars, what poisonous trace gases might also be there, oozing out of the rocks? All of the schemes are fantastically expensive and impossible given our current technology. The impossibility of doing these things, for now, absolves us from the responsibility of deciding whether we should.

On Venus, the major impediments to our comfort are the temperature, the pressure, the unbreathable CO_2 atmosphere, and the lack of water. (Otherwise it's quite nice!) Terraforming schemes all focus on getting rid of large amounts of CO_2 and converting some of it to oxygen. This would solve three out of four problems. Then just add water and voilà! There is another problem that some of the schemers and dreamers do not consider: if Venus had Earth's atmosphere but retained its very slow rotation, daytime heat and nighttime cold might be unbearable. It's hard to say what kind of weather would result. So, while we are scheming, we should also figure out how to spin Venus up to a more "normal" rotation rate.

In brief, here are four ideas proposed to "fix" Venus to be more like Earth:

> **1.** Pummel it with asteroids and comets, blowing off a lot of the atmosphere *and* adding cometary water. A coordinated targeting could also help speed up the spin.

> **2.** Make large dust clouds in the upper atmosphere that would shut down the greenhouse effect and cool the planet. The asteroid pummeling mentioned above would help here by kicking up quite a dust cloud on its own.

---

*This may seem far out to you, but hey, Las Vegas and Tucson have been terraformed, at least temporarily, so anything is possible.

3. Use orbiting sunshades to cool off the planet enough to condense out the CO_2. Then cover up the resulting lakes with reflecting fabric, so they don't evaporate back into the air.

4. Seed the clouds with algae that convert CO_2 to O_2. Genetic engineering could provide custom bugs that enjoy life on Venus and eventually make it more like Earth.

These schemes may seem a bit fantastic, even silly. Each has been discussed in much more elaborate terms, with detailed calculations of the energies and timescales involved. This doesn't make any of them sound much more credible, though. The least offensive and the most doable is the last one. Biological solutions do not take huge amounts of energy because they use the natural, exponentially multiplying quality of life. If we found or created a suitable organism, we would need to introduce only a small amount and it would eventually spread and transform the planet.

One more possibility is worth mentioning. In Chapter 5 I pointed out that some new evolutionary climate models hint at instability in the climate of Venus, or at least the potential for instability that we might be able to trigger. If so, we might use a minor application of any of the above schemes to jump-start a runaway cooling. Cool things down just a little, and surface-atmosphere chemical reactions might remove CO_2 from the atmosphere, reducing the greenhouse effect and cooling things still further. The resulting feedback loop would cause the climate to decay to a condition we would find more comfortable.

These schemes seem impossible and perhaps unethical, so why even bother to discuss them? Well, today's impossibility is tomorrow's engineering project. Who would have thought, one hundred years ago, that we would have walked on the Moon (and then abandoned it!) by now? The dreamers are sometimes the advance wave of the next revolution. But there is a much better reason to be talking about this now. Thinking about terraforming is a good intellectual sport. It's fun, and it stretches our mind muscles in *useful* ways. Fantasizing about how we would terraform other planets helps us think about how planets and climates work and how they might change. Thinking about changing planets *in a purposeful way* is especially good for us, because we have been changing our own planet in random, haphazard ways. We need to learn how to take a more collective, constructive approach to altering our planet's environment. By imagining how we could do this elsewhere, we enlarge our understanding of global change. Any successful terraforming project would take generations to complete. It's good for us to practice thinking about global change on such long timescales. In fact, our survival will depend on our learning how to do this. The value of thinking about terraforming is that it can help us learn how to stop *veneraforming* our own planet (making it more like Venus).*

Understandably, talk of monkeying around with planets makes many people uncomfortable. There is a natural resistance to such ideas, given our sorry recent history of tinkering with this world beyond our ability to foresee or cope with the consequences. There is a strong reaction against the human arrogance of thinking we can "play God" with the solar system when we seem barely able to manage our affairs here. We do have a history of charging off and "exploring" places in an intrusive and impulsive way, messing things up, then realizing it later and issuing half-hearted apologies. We must learn from this, and step mindfully into the solar system. At the very least, we should hold ourselves to the standard of being able to consciously terraform Earth (rather than unconsciously veneraforming it as we are doing now) before we take on other worlds.

Finally, there is a further moral dimension: Do we know for sure there is not extant life on Venus or Mars that we would be destroying? We must be much more certain about this than we are now before we risk, in our ignorance, wiping out a biosphere.

Even so, we may one day decide that it is a good idea to transform other worlds. If we ever determine, *for sure*, that, say, Mars is utterly lifeless (something not possible with our current scientific abilities), then the ethical imperative changes. Propagating life is arguably a good thing, and *preserving* life is certainly a good thing. If we want to ensure that Earth life, including humans and other creatures, survives the planetary disaster that surely will come someday (by asteroid, comet, human stupidity, or something unanticipated), we must gain permanent outposts beyond Earth. Mars and Venus may eventually be options for this kind of transplanetary life insurance policy.

*See "Venus Envy" in Chapter 4.

We are community, within and without.* It's only natural to want to extend this and create an interplanetary community. There is beauty and inspiration in the vision of humanity's spreading into the galaxy, leaving the cradle, becoming who-knows-what. Throughout my teenage years, this was one of the guiding visions of my life. I have since gained a healthy respect for our current limitations. I still believe in the vision, but I think we have some major tests to pass first. We have to do our homework before we can graduate to being a wise, space-faring species. To survive, thrive, and possibly expand, we must learn some of the lessons Earth is trying to teach us right now—lessons about our limits and our interconnectedness, lessons about what it means to function and live sustainably as part of a planetary community.

Fortunately, it will be a long time before we are capable of terraforming, so we can mull over the ethical implications for at least a century and perhaps several. Maybe we will decide that it is not such a good idea. Terraforming will remain a purely intellectual exercise for the foreseeable future and, as such, it is fun and arguably worthwhile to fantasize about. But anyone who suggests seriously that we embark on any of these schemes anytime soon should be institutionalized or forced to teach Freshman Astronomy at a large public university. Any actual plan to implement such changes would be the height of folly until we learn *much more* about life, about planets, and about ourselves.

OUR PLANET, OURSELVES

Space ain't man's final frontier. Man's final frontier is the human soul.
—ARRESTED DEVELOPMENT, "Man's Final Frontier"

Remember, the mountains don't care.
—SIGN AT TRAILHEAD, Rocky Mountain National Park

In earlier times Venus was an object with great spiritual significance. Hundreds of years ago people in Mesoamerica and many other places saw it as an animated presence closely connected to the underworld and to human origins. In the last four hundred years we have seen Venus take on many personas, heavenly and hellish. The ideas and models we have used to describe her reflect changes in our views of

*That is, our bodies are, in a sense, communities of microorganisms, and our biosphere is an intricate mesh of interacting communities.

ourselves, our relationship with our world, and our relationship—once we knew there was one—with the larger universe. Now we find that in some important ways Venus is a long-lost sister to Earth, a nearby world that is complex and active like our own. This is no demotion in status: being a twin to our home planet is not a lesser thing than being a god.

The human race and the planet Earth are at a crossroads now. Our recent spurt of new knowledge about Venus comes at a time when we are once again reevaluating our relationship with our planet. From one perspective, Earth was doing fine before we came along and will be better off if we do ourselves in. But this view overlooks the unique and precious attributes that we, with all our failings, bring to the planet. After all, we are, as the song goes, the eyes of the world, as well as the brain, heart, mind, and soul. Surely the world would hum along without us and perhaps be grateful for the end of the ruckus, but it may be left in a brain-dead state, especially if we take the dolphins and chimps out with us. I don't mean to suggest the conceited notion that we are the only intelligent species that will ever inhabit this corner of the universe. Given a little time, someone else will find it useful to become self-aware. Perhaps elephant herds or termite mounds, given a few million years, will surprise themselves by learning to use tools and language. I would even question whether the term *intelligent* applies to us. I would suggest the following test: a species may be considered intelligent if it is capable of taking care of itself and other species, of foresight and collective action, and especially of not overpopulating and soiling its own pen to the point of extinction. By this definition it is not at all clear that intelligence has yet evolved on Earth. Perhaps that explains why no one has answered our first radio messages to the cosmos, our interstellar Chuck Berry records, or even the reruns of *My Favorite Martian* that have now spread tens of light-years into the galaxy. Human-bashing is one popular response to our current dilemmas. But right now we are the only game in town, so what can we do but try to survive and act "intelligently"?

We face the self-imposed challenges of overpopulation, resource depletion, global warming, ozone destruction, and the slow spread of chemical poisons known and unknown. The realization is growing that we need to change the way we operate within our planetary home. Various individuals and groups have their schemes, but there is widespread disagreement, disillusionment, cynicism, and skepticism. It's a frightening situation: we know we need to change the way we do business in order to survive, but we don't know how.

As a species, we are going through a painful adolescence. We are self-aware but not yet in control. We have many newfound powers but haven't given enough thought to what we should be doing with them. Somewhat aware of the consequences of our actions, we are not yet willing and able to take responsibility for ourselves. We are hooked on immediate gratification and not used to cleaning up our messes. We have little awareness of limits, and we love to watch things blow up. While we enjoyed the naive, infinite resiliency of youth, we could afford to be oblivious, but we are pushing the limits of our planet's capacity.

Adolescence is not all bad. A tremendous creativity and boundless exuberance makes the teenage years a lot of fun and makes teens so great to be around. We must harness that energy for survival rather than suppress it. But how does maturity come? Partially through mentoring. It would be nice if some aliens who had already passed through this difficult phase would come and help us get a grip.* Unfortunately, we cannot count on this. We had better do it ourselves.

Maturity is also aided by an expanded reference frame, including knowledge of other places and new perspective on oneself. Perhaps the heightened planetary self-awareness provided by human encounters with other worlds can help us rise to this challenge.

We have heard many variations on the saying, "They can put a man on the Moon, but they can't [insert phrase here]." (My favorite is the funny/feminist ending "Why can't they put them all there?") It's true: we put several men on the Moon, but we can't put food on everyone's table, stop fighting, and learn to clean up after ourselves. Clearly science alone will not solve our problems. Some new technological tricks may help, but what we really need is the mass acceptance of new ways of thinking about ourselves. We need a widely credible and hopeful vision of our future.

The main obstacles to our survival at this point are not technical and scientific but sociopolitical and spiritual. In reaching out across interplanetary space to explore our sibling worlds, however, we need not separate scientific and spiritual quests. Earth is our body, our

*This is one of the themes of A. C. Clarke's *Childhood's End.*

home, and thus physical knowledge of planetary history *is* spiritual knowledge of self.

Is it really in our nature to destroy ourselves, to ruin Earth for our kind of life? Or are we just stuck for the moment, lacking a vision of how to proceed safely into the future? We have taken on and created for ourselves some large and very new challenges. Our biggest need may just be to look at our situation differently.

All this "collective" language is nice, but in what sense do we act as a species? We make decisions as individuals, groups, governments, and corporations. How can we hope to solve problems that require a change in the behavior of humanity as a whole when we have no mechanisms to act collectively? The planetary perspective on who and where we are can help us here. A wonderful example is the depletion of Earth's ozone layer. We recognized this mistake through comparative planetology, and we seem to be on a course toward correcting it through planetwide agreements. Planetary consciousness breeds planetary responsibility.

When we step back and reflect on what we have learned, the most important benefit of planetary exploration will be self-knowledge. Venus *is* our twin, not identical, but recognizable, and full of the promise of enlarged perspective for Earth and its newly conscious residents. We should treasure every bit of knowledge and insight Venus can provide. It's the only twin we've got.

Fractured plains, 235, 235f
Frankenstein (Shelley), 52n

Gabor, Zsa Zsa, 53, 54f, 63
Gaia hypothesis, 154–157
    and theories of life, 294–296, 302
Galilei, Galileo, xiv, 32, 33–34,
        35–37, 62, 170, 218
Galileo, 124
    spectroscopy, 133
Ganymede, 25
Genesis, 179
Geology, focus of, 197
Geophysiology, 294, 305
Goldstone radio dish, 202, 213
Gradient, defined, 101, 297
Granite, 230–231
Greeks, astronomy among, 33
Greenhouse effect, 145, 146f, 147,
        163, 263
    feedbacks affecting, 152–154
    runaway, 149
Greenwich Observatory, 206
Guatemala, 19
Guinevere Planitia, 235f

Hadley cells, 141, 142f, 277
Hadley circulation, 159
Hale-Bopp comet, 112n
Halley's comet, 110, 112f, 137
Harjo, Joy, 138
Hawaiian Islands, 175
Haxti, 21
Hell, notions of, 70–71, 70n, 71n
Hendrix, Jimi, 71
Hermaphrodite, etymology of, 24f
Herschel, William, 39
Herzog, Werner, 286
Hesperos, 25
Himalaya, 166
Hitler, Adolf, 63
Honeybees, 42n
Hot spots, 174, 176f
    land creation by, 176–177

Hot-spot-cold-spot controversy,
    234
Hoyle, Fred, 50
Hubble Space Telescope, 123
    flaws in, 215
    images from, 180f
    value of, 179
Hunter, Robert, xi, 83, 126
Hutton, James, 184, 185
Huygens, Christiaan, 38
Hyakutake comet, 112
Hydrochloric acid, 317
Hydrogen
    heavy, 106–108
    prevalence of, 99n
    reactivity of, 99

Iapetus, 44
Impact basin, defined, 91n
Impact craters, 252
    on Earth, 188, 188f
    partial burial of, 257f
    on Venus, 253–262, 255f, 256f
Impact-extinction hypothesis, 187,
    189
    fallout from, 191
Inanna, 18
Inferior planets, 5
Intermittent plate tectonics,
    266–267
International Astronomical Union
        Committee on Nomenclature,
        26
Interplanetary space, 64
Inversion, temperature, 160
Iron sulfide, 246–247
Ishtar, 17, 18, 25
Ishtar Terra, 207f, 209, 232, 235,
    239–243
Isotopes, 115–116

Janus, 60n
Jefferson, Thomas, 186
Jodrell Bank radiotelescope, 76